IGA: Non-conforming Coupling and Shape Optimization of Complex Multipatch Structures

Isogeometric Analysis Tools for Optimization Applications in Structural Mechanics Set

coordinated by
Piotr Breitkopf

Volume 1

IGA: Non-conforming Coupling and Shape Optimization of Complex Multipatch Structures

Robin Bouclier
Thibault Hirschler

iSTE

WILEY

First published 2022 in Great Britain and the United States by ISTE Ltd and John Wiley & Sons, Inc.

ISTE Ltd
27-37 St George's Road
London SW19 4EU
UK

www.iste.co.uk

John Wiley & Sons, Inc.
111 River Street
Hoboken, NJ 07030
USA

www.wiley.com

Library of Congress Control Number: 2022931519

British Library Cataloguing-in-Publication Data
A CIP record for this book is available from the British Library
ISBN 978-1-78630-824-5

Contents

Preface

P.1. The book series

This book constitutes the first volume of the set of books entitled "IsoGeometric Analysis (IGA) tools for optimization applications in structural mechanics". The objective of the series is to present, in a comprehensive and detailed manner, some advanced modeling and numerical strategies, recently developed in the context of *IGA*, that provide strong benefits not only for *direct* simulations but also for the resolution of *optimization* problems in *structural mechanics*. The IGA paradigm was originally introduced by Hughes and co-workers in 2005 (Hughes *et al.* 2005) to reunify the fields of computer-aided design (CAD) and finite element analysis (FEA). The core idea is to resort to the same higher-order and smooth spline bases for the representation of the geometry in CAD as well as for the approximation of solutions fields in FEA. The use of such families of functions quickly made IGA highly attractive for two main reasons: first, a common geometrical model can be used by both the designers and analysts; second, an increased per-degree-of-freedom (per-DOF) accuracy can be reached in comparison to standard finite element methods (FEM). This technology is thus often seen as a high-performance computational tool.

Beyond its undeniable superior analysis properties, IGA is also highly relevant for addressing the higher-level problems of optimization (also characterized as inverse problems) that enable us to question and adapt a numerical model with regard to key features. Indeed, IGA provides a *natural regularization* framework for such problems since it allows us to look for the solution in more regular approximation subspaces. Following this mindset, many progresses can be reported in the current literature, with the common goal of further consolidating IGA by demonstrating its performance for optimization-type applications. This book series is meant to be part of this attempt.

P.2. The present volume

This first volume tackles design optimization that is essential to reduce the usual long trial-and-error learning process during the product development in industry. More precisely, we focus on *structural shape optimization*, which aims at finding the optimal shape of a given topology of a structure with respect to a certain objective, such as minimal mass and maximal rigidity. Shape optimization undoubtedly constitutes one of the most valuable applications for IGA as the latter combines high-quality geometric representations and efficient analysis capabilities into one single model. In this book, we seek to develop a *full analysis-to-optimization framework* that is applicable to *complex structures* in order to tend to real-world (industrial) applications.

From an analysis point of view, this forces us to face two very important challenges encountered in the field today, namely (i) the simple local enrichment of rigid tensor-product spline models (which may also go with trimming when introducing an arbitrary local region within a spline patch) and (ii) the efficient analysis of large multipatch structures. The methods proposed to answer these issues are quite original for the domain and are based on developments in which the French community is particularly active. More precisely, the strategies rely on domain coupling and, in particular: (i) *global/local non-invasive coupling algorithms* that enable us to locally enrich a global model without altering its corresponding numerical operators and (ii) *parallel domain decomposition solvers* that allow us to separate the computational resources between several subdomains.

In a next step, the developed direct solvers are finely integrated into the shape optimization loop for full robustness and efficiency, and a *geometrical modeling*, based on embedded entities, is formulated to arrive at what we may refer to as *optimization-suitable* models, i.e. models that are ready for natural isogeometric shape optimization regardless of their complexity. The resulting framework is very robust and generic; it merges all the attractive features of IGA (design–analysis link, numerical efficiency, natural regularization, etc.)

and thus brings the possibility of exploring new types of design. The performance of the methodology is demonstrated in the context of elastic solid and thin shell structures (with the Kirchhoff–Love theory). In particular, we consider as a guiding application the optimal design of stiffened structures, which are ubiquitous in aeronautics. Today, designs of wings and fuselages contain a lot of straight and aligned stiffeners uniformly distributed. However, it seems that defining curvilinear stiffeners instead of straight ones can further improve the mechanical behavior of the overall stiffened structures. Figure P.1 highlights what could be, in the future, the aerostructures based on free-form stiffeners. The developed framework contributes to reach such innovative designs.

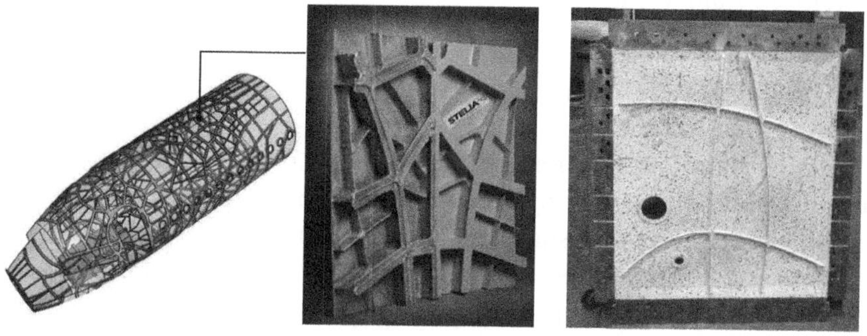

(a) (b)

Figure P.1. *Innovative stiffened structures with curvilinear stiffeners: (a) from Renard (2018) and (b) from Havens* et al. *(2011). For a color version of this figure, see www.iste.co.uk/bouclier/IGA.zip*

P.2.1. *Intended audience*

Overall, this book provides a contemporary vision of IGA by first discussing the current challenges to achieve a true CAD-analysis bridge, then it proposes original solutions to answer the related issues from an analysis point of view and eventually goes up to the shape optimization of structures, which is one of the greatest applications of IGA. This book is intended to be sufficiently self-contained so as to not require that much background in IGA. We wish this book to be accessible for computational scientists, with a good background on finite element analysis and structural mechanics, and to be addressed to people either already part of or willing to join the IGA community. Throughout the chapters, we endeavor to provide detailed information and accurate datasets so that the reader should be able to understand all the essential ideas and to reproduce the numerical experiments. In addition, an exhaustive list of relevant references is provided as it is not possible to address every topic in the full generality and completeness that it deserves. In particular, the first and leading book in IGA by Cottrell *et al.* (2009), as well as the research papers by the authors upon which most of the content is based (see, for example, Bouclier *et al.* (2016, 2017), Bouclier and Passieux (2018), Hirschler *et al.* (2019a, 2019b, 2019c)), obviously constitute relevant additional resources. Finally, conscious effort has been made to present material that is *not* in research papers and to draw relevant perspectives, which in our opinion allows us to appreciate the full potential of the developed framework.

P.2.2. *Organization of the text*

Following these opening remarks, this book is organized as follows: first, Chapter 1 re-introduces IGA by highlighting the opportunities and remaining issues for the analysis and shape optimization of complex structures; then, Chapters 2 and 3 are devoted to the derivation of the direct solvers, i.e. the non-invasive coupling scheme for flexible global/local simulation and the family of parallel domain decomposition solvers for efficient multipatch analysis, respectively; finally, the constructed algorithms are integrated into the shape optimization loop and complemented by a sound geometrical modeling to achieve the optimal design of innovative complex structures in Chapter 4. From here on, it should be stressed that the contributions regarding optimization concern the modeling and the involved direct resolutions rather than the optimization algorithm itself.

The first chapter serves as a prerequisite for the others. It does not include innovative developments, but is intended to provide all the necessary information on IGA by discussing in particular the current challenges related to the true dialog between CAD and FEA. This chapter may be helpful for the reader who is unfamiliar with IGA and helps explain the motivations of our work. Although Chapters 2 and 3 share many similarities from the conceptual point of view and both start with domain coupling, they can be read independently according to the reader's interest. Chapter 4 can also more or less be read on its own by people involved in IGA; yet, fully skipping Chapters 2 and 3 would inherently lead to difficulties in clearly understanding the benefits of the non-invasive and domain decomposition solvers to be integrated in shape optimization loops. All chapters, except Chapter 1, start with a precise introduction to pave the context and end with a conclusion that summarizes our most important points and motivates future research based on the proposed methodology.

P.3. Acknowledgments

The book series has been prepared under the suggestion of Piotr Breitkopf, director of the ISTE series "Numerical Methods in Engineering", following Robin Bouclier's (2020) thesis, known as a "Habilitation à Diriger des Recherches" (HDR). Many thanks to Piotr for giving the authors this opportunity and for helpful comments and advice concerning an initial draft of this book. The authors would also like to thank their collaborators on the work contained in this volume. In particular, this volume completes many texts and results from the PhD thesis of Thibaut Hirschler (2019); thus, the authors would like to gratefully thank the colleagues from the supervision team of this PhD, starting with Thomas Elguedj and going up to Joseph Morlier without omitting Arnaud Duval. Finally, Robin Bouclier would like to single out for special acknowledgments Jean-Charles Passieux and Michel Salaün who initiated him into the field of domain coupling, specifically in the context of non-invasive global/local simulations, at the early stage of his arrival in Toulouse (France).

Robin BOUCLIER
Thibaut HIRSCHLER
January 2022

1

Introduction to IGA: Key Ingredients for the Analysis and Optimization of Complex Structures

1.1. Brief introduction

IsoGeometric analysis (IGA) was originally introduced by Hughes *et al.* (2005) and formalized in Cottrell *et al.* (2009), in order to reunify the fields of geometric modeling in computer-aided design (CAD) and numerical simulation using the finite element method (FEM). The main idea is to resort to the same bases for analysis as the ones used to describe the geometry in CAD. In this framework, the method can be viewed as a generalization of the FEM that considers smooth and higher-order functions, for example, the non-uniform-rational-B-spline (NURBS) functions (Cohen *et al.* 1980; Piegl and Tiller 1997; Rogers 2000; Farin 2002), to replace typical Lagrange polynomials in the computations. Some other geometric descriptions include T-splines (Bazilevs *et al.* 2010) and subdivision surfaces (Cirak *et al.* 2002). Within this work, we only use the NURBS (which constitute the most commonly used technology in CAD) and simpler B-splines. We use the *spline* and *isogeometric* terminologies indifferently to denote a NURBS and a B-spline object, respectively.

Now, about 15 years after its birth, there are substantial works in the area of IGA, which makes it a very competitive methodology for the general field of scientific computing. In this chapter, we attempt to introduce IGA by providing a contemporary vision on its interests, limitations and related challenges that still need to be faced in order to meet its full potential. The discussion is performed in accordance with the core objective of this book, which is the *analysis* and *shape optimization* of *complex structures*. More precisely, we insist on the opportunities offered by the underlying spline technologies to represent, modify and mechanically simulate any geometrical shape in structural mechanics. We also outline the remaining issues, and the main research paths currently followed to answer them, to achieve a true dialog between CAD and FE analysis. An effort is made to help readers unfamiliar with IGA to understand all of the necessary key points related to this method. This first chapter serves as a prerequisite for the contributions presented in the next chapters of this book.

1.2. Geometric modeling and simulation with splines

To start with, IGA is introduced from a technical viewpoint by providing the key ingredients regarding the considered spline geometric modeling techniques, namely the B-spline and NURBS variants. Particular care is taken to highlight the ability of these spline tools to describe any geometrical shape and to also control them smoothly. These aspects are of paramount importance for the works presented in this book, which address the general field of computational solid mechanics to the shape optimization of structures. Finally, with the IGA concept now being mature and relatively well known in the scientific computing community, we shortly review its basics from an analytical point of view. In this respect, we recall that the major difference, with respect to standard FEM, is to use the spline-based parameterizations of CAD to build the approximation subspaces when applying the Galerkin's method. For further details, besides the pioneering contributions (Hughes *et al.* 2005; Cottrell *et al.* 2009), refer to the works cited hereafter.

1.2.1. *Parametric representation of geometries*

The spline formalism offers a natural way to represent geometries in the parametric form. Hence, let us specify the parametric representation of geometries before entering into the details of the B-spline and NURBS technologies. To begin with, we consider univariate geometric entities living in 2D physical spaces. These entities simply consist of planar curves. Describing these curves in the parametric form consists of expressing each coordinate of a point on the curve separately as an explicit function of an independent parameter (Piegl and Tiller 1997):

$$C(\xi_1) = \big(x(\xi_1),\, y(\xi_1)\big), \quad \forall \xi_1 \in \mathcal{I}_1, \qquad [1.1]$$

where $\mathcal{I}_1$ is the parameter space (usually taken as the normalized interval $[0, 1]$), i.e. that defines the space of variation of parameter ξ_1. As examples, we provide the parametric descriptions of the line and the circle depicted in Figure 1.1. The line is defined as:

$$x(\xi_1) = \xi_1 \quad \text{and} \quad y(\xi_1) = -\frac{b}{x_0}\xi_1 + b. \qquad [1.2]$$

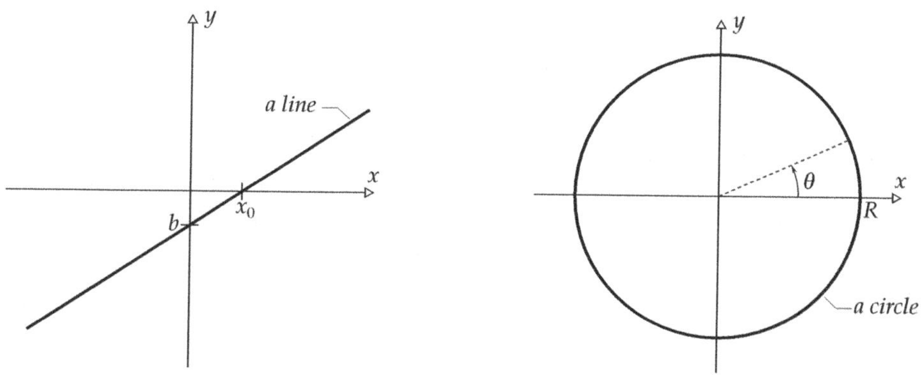

Figure 1.1. *Two simple planar curves*

Parameter ξ_1 takes any real value to represent the complete line or is bounded to represent a segment of the line. Using the same idea, the curve of the circle can be parameterized as follows:

$$x(\theta) = R\cos\theta \quad \text{and} \quad y(\theta) = R\sin\theta, \qquad \forall \theta \in [0, 2\pi]. \qquad [1.3]$$

Note that at this stage the parameterization is not unique and parametric forms are obviously not the only way of representing curves. Implicit equations are also commonly used to describe curves. For example, the circle of radius R and centered at the origin can be described by:

$$x^2 + y^2 - R^2 = 0. \qquad [1.4]$$

Both techniques have different advantages. The parametric form gives a direct way to compute a point on a curve. On the other hand, with the implicit form, it is easy to determine if a given point is on the curve or not.

The previous concept can be straightforwardly extended to multivariate geometric entities living in 2D or 3D physical spaces. There exist typical examples like those represented in Figure 1.2. For example, the unit surface sphere can be parameterized as:

$$S(\theta,\varphi) = \big(x(\theta,\varphi),\, y(\theta,\varphi),\, z(\theta,\varphi)\big), \tag{1.5}$$

with:

$$\begin{aligned} x(\theta,\varphi) &= \sin\theta\cos\varphi, \\ y(\theta,\varphi) &= \sin\theta\sin\varphi, \quad \forall\theta\in[0,\pi],\ \forall\varphi\in[0,2\pi]. \\ z(\theta,\varphi) &= \cos\theta. \end{aligned} \tag{1.6}$$

Two parameters are needed to describe a surface (θ and φ for the sphere). For describing a volume, three parameters are required. For example, the hollow cylinder from Figure 1.2 is described by:

$$V(r,\theta,h) = \big(x(r,\theta,h),\, y(r,\theta,h),\, z(r,\theta,h)\big), \tag{1.7}$$

with

$$\begin{aligned} x(r,\theta,h) &= r\cos\theta, \\ y(r,\theta,h) &= r\sin\theta, \quad \forall r\in[R_i,R_e],\ \forall\theta\in[0,2\pi],\ \forall h\in[-H/2,H/2]. \\ z(r,\theta,h) &= h. \end{aligned} \tag{1.8}$$

NOTE.– Finally, note that the direct description of geometric entities in the parametric form may not be possible when these differ topologically from a square or a cube. In this case, Boolean operations between different parametric entities may be required to intersect and/or join them. We will see that this issue is reflected in IGA: it leads to facing non-conforming multipatch analysis, which is one of the current important challenges in the field and is largely addressed in this book.

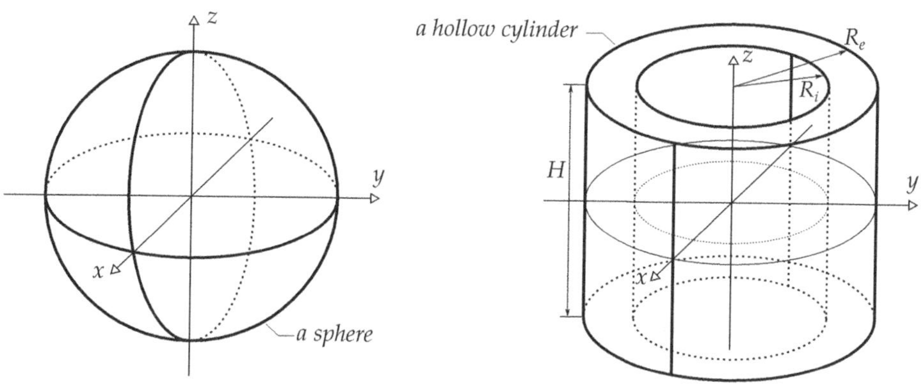

Figure 1.2. *Two simple (surface and volume) geometric objects*

1.2.2. *B-spline and NURBS technologies*

The B-spline and NURBS families are the spline technologies that have become the standard for geometric modeling in CAD and computer graphics (Cohen *et al.* 1980, 2001; Piegl and Tiller 1997; Rogers 2000; Farin 2002) over the years. The NURBS functions lend themselves to an exact representation of many shapes used in engineering, such as conical sections (circles, cylinders, spheres, ellipsoids, etc.). NURBS are a generalization of B-splines: they can be viewed as rational projections of B-splines that are piecewise polynomials. Therefore, they possess many of the properties of B-splines, the most interesting one being their possible increased smoothness, thus implying models with few degrees of freedom. Since NURBS are based on B-splines, we start the presentation below with B-splines.

1.2.2.1. *B-spline curves*

The B-spline technology enables us to describe geometric entities in terms of smooth piecewise polynomials (Gordon and Riesenfeld 1974; Piegl and Tiller 1997). First, taking the univariate case, it expresses curves in the parametric form as:

$$C^h_{BS}(\xi_1) = \sum_{i=1}^{n_1} M^p_i(\xi_1) P_i, \quad \forall \xi_1 \in \mathcal{I}_1, \tag{1.9}$$

where P_i is the control point associated with the univariate pth-degree B-spline basis function M^p_i. Note that P_i can live in a 1D ($P_i = x_i$), 2D

($P_i = (x_i, y_i)$) or 3D ($P_i = (x_i, y_i, z_i)$) physical space. As stated above, the n_1 B-spline basis functions are piecewise polynomial functions. Their expression is obtained by specifying a *knot-vector* Ξ_1 and a polynomial degree p. Thus, three inputs are required to define B-spline curves: the degree, the knot-vector and the control points. The degree p obviously takes an integer value. The knot-vector $\Xi_1 = \{\xi_1^1, \xi_1^2, .., \xi_1^{n_1+p+1}\}$ is a non-decreasing sequence of $n_1 + p + 1$ real numbers, referred to as *knots*, that divide the parameter space into *knot-span* elements. The interval $\left[\xi_1^1, \xi_1^{n_1+p+1}\right]$ constitutes the spline *patch*. Unlike standard FE where each element has its own parameterization, the parameter space of B-spline functions is localized onto the patch that may be thought of as a macro-element. Preliminary illustrations of B-spline entities, along with their B-spline functions, are provided in Figures 1.3 and 1.4. Usually in IGA, the first $p + 1$ knots are identical and equal to the parameter lower bound, and the last $p + 1$ knots are identical and equal to the parameter upper bound. In this case, the basis is interpolating at the boundary knots of the patch, which facilitates the application of boundary conditions when performing numerical simulations. This treatment is performed in this work. Considering $\mathcal{I}_1 = [a_1, b_1]$, this leads to so-called *open* knot-vectors that are of the form:

$$\Xi_1 = \{\underbrace{a_1, \ldots, a_1}_{p+1}, \xi_1^{p+2}, \ldots, \xi_1^{n}, \underbrace{b_1, \ldots, b_1}_{p+1}\}, \qquad [1.10]$$

where the interior knots verify the conditions $\xi_1^i \leq \xi_1^{i+1}$ and $a_1 < \xi_1^i < b_1$.

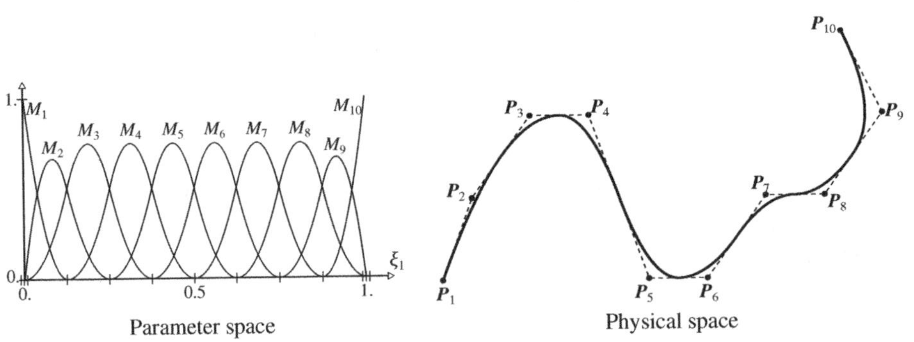

Figure 1.3. *Quadratic B-spline curve with* $\Xi_1 = \{0, 0, 0, \frac{1}{8}, \frac{2}{8}, \ldots, \frac{7}{8}, 1, 1, 1\}$

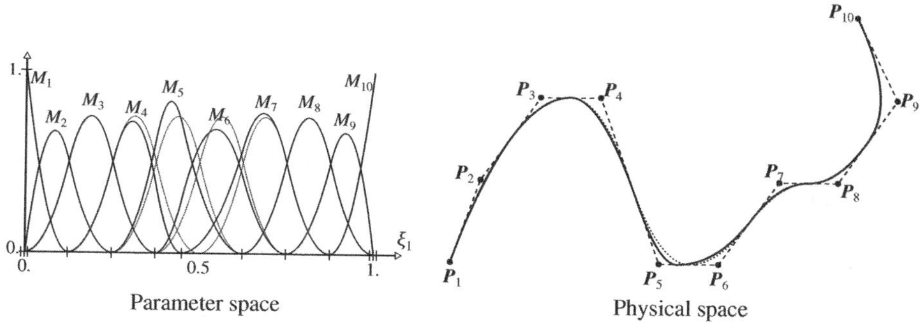

Figure 1.4. *Quadratic B-spline curve with a non-uniform knot-vector. The initial knot $\xi_1^4 = \frac{4}{8}$ used in Figure 1.3 has been changed by $\xi_1^4 = \frac{9}{20}$ (the dotted lines repeat the uniform knot-vector case)*

Returning to the examples, Figure 1.3 depicts a quadratic (i.e. $p = 2$) B-spline curve, where the knot-vector was chosen as:

$$\Xi_1 = \{0, 0, 0, \tfrac{1}{8}, \tfrac{2}{8}, \tfrac{3}{8}, \tfrac{4}{8}, \tfrac{5}{8}, \tfrac{6}{8}, \tfrac{7}{8}, 1, 1, 1\}. \quad [1.11]$$

This knot-vector is characterized as *uniform* since all interior knots are equally spaced. The influence of the knot-vector is shown in Figure 1.4, where we only changed one knot of the initial knot-vector such that:

$$\Xi_1^{\text{new}} = \{0, 0, 0, \tfrac{1}{8}, \tfrac{2}{8}, \tfrac{3}{8}, \tfrac{9}{20}, \tfrac{5}{8}, \tfrac{6}{8}, \tfrac{7}{8}, 1, 1, 1\}. \quad [1.12]$$

This new knot-vector is characterized as *non-uniform* because the interior knots are not equally spaced.

1.2.2.2. *B-spline basis functions*

So far, we have not mentioned how the B-spline basis functions involved in equation [1.9], and already plotted in Figures 1.3 and 1.4, are formulated. There are different ways to define them (Piegl and Tiller 1997). The standard practice in IGA is to call on the following recursive formula:

$$M_i^0(\xi_1) = \begin{cases} 1 & \text{if } \xi_1^i \leq \xi_1 < \xi_1^{i+1} \\ 0 & \text{otherwise} \end{cases},$$

$$M_i^p(\xi_1) = \frac{\xi_1 - \xi_1^i}{\xi_1^{i+p} - \xi_1^i} M_i^{p-1}(\xi_1) + \frac{\xi_1^{i+p+1} - \xi_1}{\xi_1^{i+p+1} - \xi_1^{i+1}} M_{i+1}^{p-1}(\xi_1). \quad [1.13]$$

This recurrence procedure is often known as the Cox–de Boor formula in reference to its inventors (C. de Boor and M.G. Cox; see Piegl and Tiller (1997) for more information). The initial functions M_i^0 are step functions. The function M_i^p is expressed as a linear combination of two $(p-1)$-degree basis functions. Benefiting from this recursive formula, efficient algorithms can be implemented to build B-spline functions (Piegl and Tiller 1997).

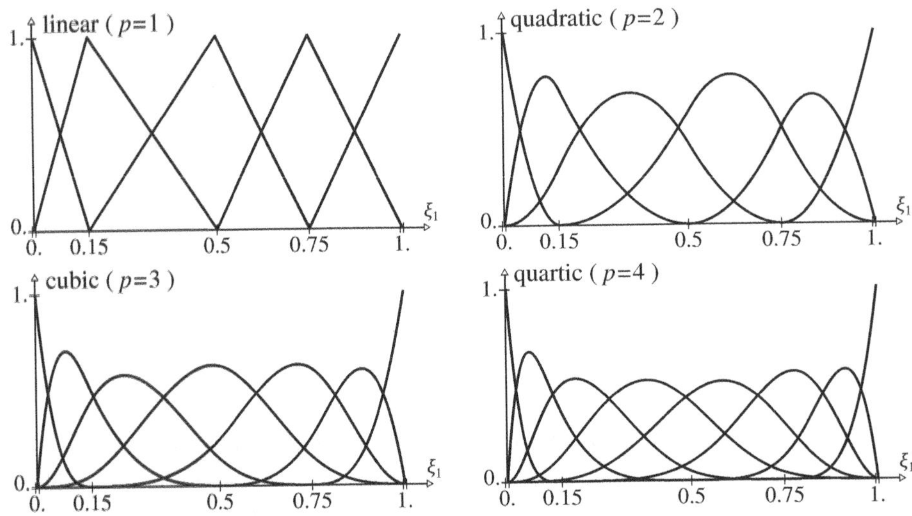

Figure 1.5. *Examples of B-spline basis functions of different degrees. The same interior knots were used (0.15, 0.50 and 0.75)*

For illustration purposes, we display several examples of B-spline basis functions in Figure 1.5. We can observe the influence of the degree. Many properties of these B-spline basis functions are given in pioneering works (Cohen *et al.* 1980, 2001; Piegl and Tiller 1997; Farin 2002). In particular, an interesting feature of B-splines is their higher degree of regularity. Indeed, a B-spline function of order p can reach a C^{p-1} regularity at knot ξ_1^i, if this one is single. There can be more than one knot at a given location of the parameter space. More precisely, if m is the multiplicity of a given knot, the functions are C^{p-m} continuous at that location, which is in contrast to standard FE, where only a C^0 regularity is encountered on the element boundaries. Figure 1.6 highlights this point: the second derivatives of the cubic basis functions are discontinuous at the double knot, while they remain continuous at single knots.

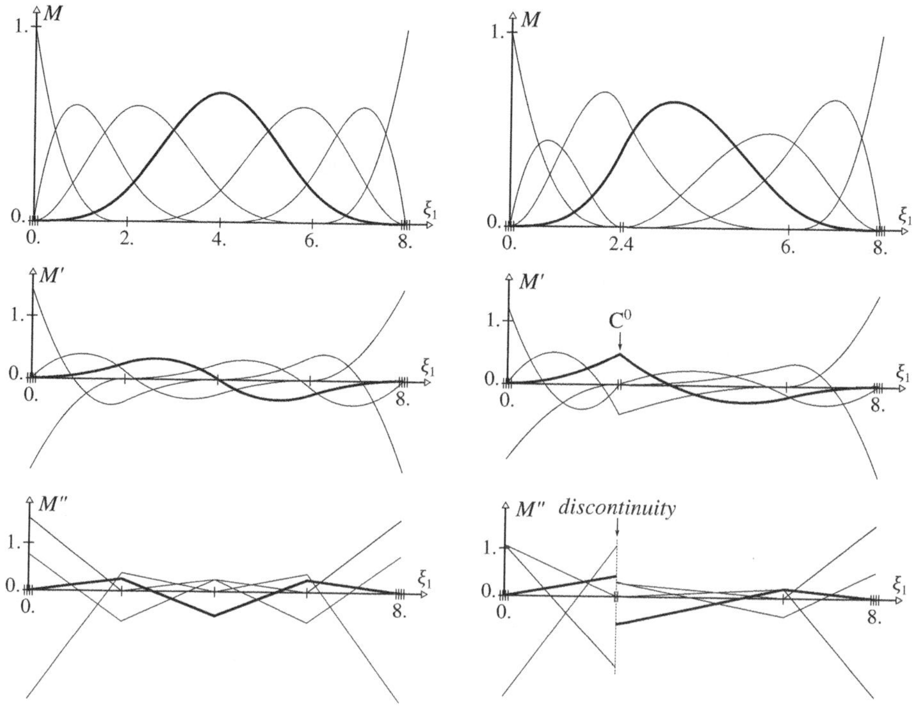

Figure 1.6. *Differentiability of cubic B-spline functions in the case of a uniform knot-vector (interior knots: 2, 4 and 6) (left), and in the case of a double knot (interior knots: 2.4, 2.4 and 6) (right)*

Last but not least, it may be noted that the computation of the kth derivatives of a B-spline basis can be efficiently performed, similarly to that for building the basis functions themselves using [1.13]. Indeed, the derivatives of B-spline basis functions are represented in terms of B-spline lower-degree bases:

$$\frac{d^k}{d{\xi_1}^k} M_i^p(\xi_1) = \frac{p}{\xi_1^{i+p} - \xi_1^i} \left(\frac{d^{k-1}}{d{\xi_1}^{k-1}} M_i^{p-1}(\xi_1) \right) - \frac{p}{\xi_1^{i+p+1} - \xi_1^{i+1}} \left(\frac{d^{k-1}}{d\xi_1^{k-1}} M_{i+1}^{p-1}(\xi_1) \right). \qquad [1.14]$$

Again, we refer to Piegl and Tiller (1997) for efficient algorithms regarding these calculations.

NOTE.– Looking more closely at the differentiability properties of B-Spline functions, it may be noted that the B-spline technology encompasses the standard FE technology based on Lagrange polynomials. Indeed, if we seek to recover the subspace of standard Lagrange polynomials with the B-spline formalism, we only need to take a knot-vector where the multiplicity of each (interior) knot is equal to p. This would generate piecewise polynomials of degree p that are C^0 at the element boundaries, which is equivalent to taking standard Lagrange polynomials to generate the approximation subspace (see Bouclier and Passieux (2022) for more information regarding the relation between FEM and IGA and, for example, Schillinger *et al.* (2016b) and Tirvaudey (2019)). Figure 1.5 (top left) clearly illustrates this statement for the case where $p = 1$: the same basis functions as standard FEM are constructed. As a result, we will obviously consider splines of degree $p \geq 2$ and, as much as possible, interior knots of multiplicity 1, in order to benefit from the increased smoothness of this technology.

1.2.2.3. *NURBS curves*

The NURBS basis functions, denoted by R_i^p in the following, are defined from a B-spline basis of the same degree M_i^p. In order to do so, it is necessary to introduce the weights w_i for the control points x_i. More precisely, the piecewise rational functions R_i^p read as follows:

$$R_i^p(\xi_1) = \frac{M_i^p(\xi_1)w_i}{W(\xi_1)} \quad \text{with} \quad W(\xi_1) = \sum_{j=1}^{n_1} M_j^p(\xi_1)w_j. \qquad [1.15]$$

Then, NURBS curves are defined similarly to those in [1.9]:

$$C_{NS}^h(\xi_1) = \sum_{i=1}^{n_1} R_i^p(\xi_1)P_i, \quad \forall \xi_1 \in \mathcal{I}_1, \qquad [1.16]$$

Consequently, a NURBS curve takes the same inputs as a B-spline curve, i.e. a degree p, a knot-vector Ξ_1 and control points P_i, plus an additional set of n_1 weights w_i. The weights give more shape control to the designer. By changing their values, a curve can be modified, as shown in Figure 1.7, for example. In fact, the positions of the control points and the values of the associated weights can be adjusted in order to build conical sections exactly (circles, ellipses, etc.; see, for example, Cottrell *et al.* (2007, 2009)). Given equation [1.15] (and verifying that the B-spline functions satisfy the partition of unity), it may be noted that if all weights are equal, the NURBS entity turns out to be a B-spline entity.

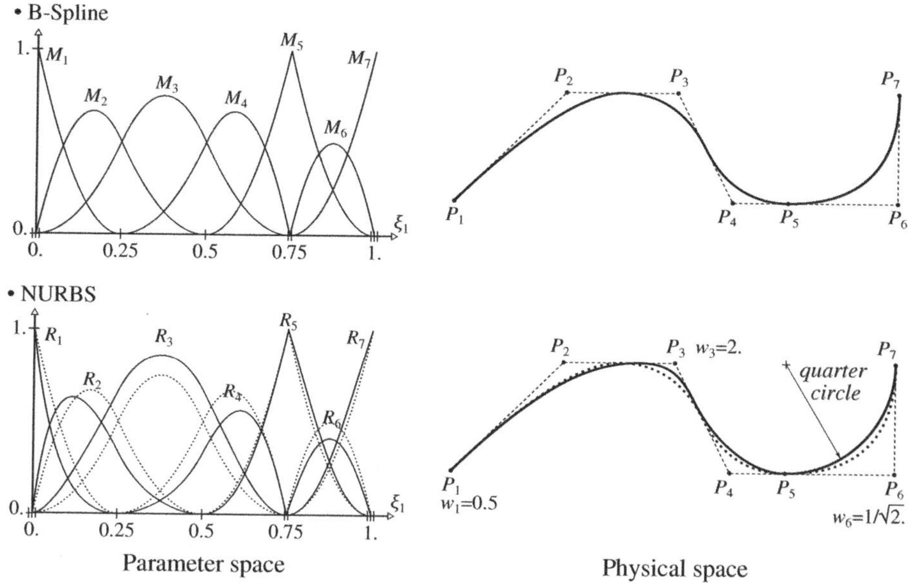

Figure 1.7. *An example of a NURBS curve: influence of the weights on the basis functions and on the final curve with comparison to the B-spline case (the weights of the first, third and sixth control points have been moved from* 1 *to* 0.5*,* 2 *and* $1/\sqrt{2}$*, respectively)*

1.2.2.4. *Surfaces and volumes*

B-splines and NURBS are not restricted to parametric curves. Obviously, multivariate entities such as surfaces and volumes can also be described in the parametric form by using these technologies. The standard construction principle is identical to that of curves. In particular, a B-spline surface is a bivariate function of the form:

$$S^h_{BS}(\xi_1, \xi_2) = \sum_{i=1}^{n_{2D}} N_i^{pq}(\xi_1, \xi_2) P_i, \quad \forall (\xi_1, \xi_2) \in \mathcal{I}_1 \times \mathcal{I}_2, \qquad [1.17]$$

where the bivariate B-spline basis functions N_i^{pq} take degree p in the ξ_1-direction and degree q in the ξ_2-direction. They are obtained by the tensor product of their univariate counterparts:

$$N_i^{pq}(\xi_1, \xi_2) = M_{i_1}^p(\xi_1) \times M_{i_2}^q(\xi_2). \qquad [1.18]$$

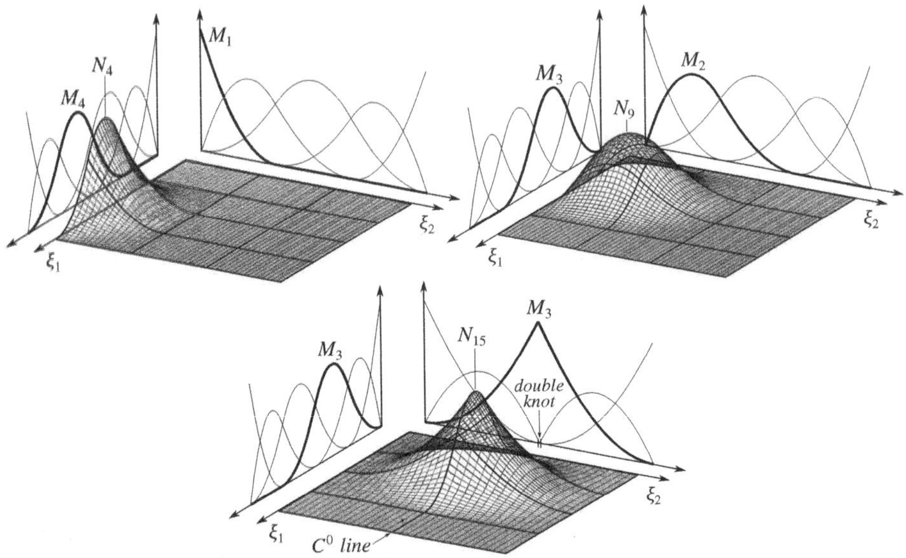

Figure 1.8. *Example of bivariate B-spline basis functions. The lower continuity at a repeated knot of the univariate basis functions propagates along a knot line due to the tensor product*

The basis functions $M^p_{i_1}(\xi_1)$ and $M^q_{i_2}(\xi_2)$ are B-spline basis functions of degree p and q and correspond to two different knot-vectors Ξ_1 and Ξ_2 of size $n_1 + p + 1$ and $n_2 + q + 1$, respectively. The number of bivariate B-spline functions is thus $n_{2D} = n_1 \times n_2$, and the ith bivariate basis function N_i^{pq} corresponds to the i_1th and i_2th functions in the ξ_1- and ξ_2-directions, respectively. Examples of bivariate B-spline basis functions are given in Figure 1.8. We can observe that the lower regularity due to a double knot leads to a complete C^0-line for the bivariate basis functions. This is due to the tensor product nature of B-spline surfaces. In other words, a pointwise feature in one parametric direction propagates along a line into the other direction. This will also make local mesh refinement delicate with such functions (see the remark in section 1.2.3). Finally, note that in order to build a B-spline surface, we should define two degrees, two knot-vectors and $n_{2D} = n_1 \times n_2$ control points (to be associated with the n_{2D} bivariate B-spline functions). An example of a B-spline surface is shown in Figure 1.9.

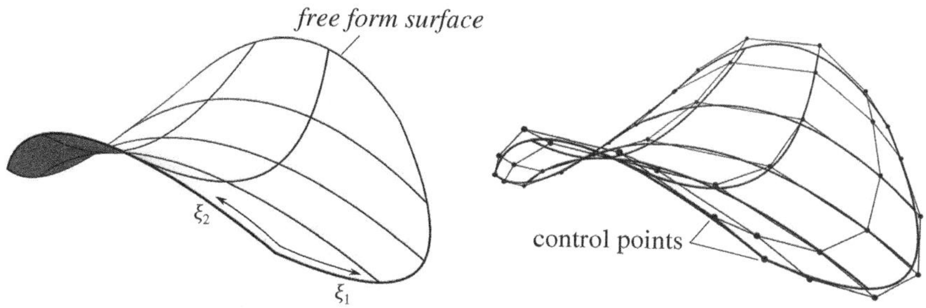

Figure 1.9. *Example of a free-form surface described with a cubic B-spline surface*

As for the univariate case, the construction of a NURBS surface requires the additional definition of one weight per control point. More specifically, a NURBS surface reads:

$$S^h_{NS}(\xi_1, \xi_2) = \sum_{i=1}^{n_{2D}} R_i^{pq}(\xi_1, \xi_2) P_i, \quad \forall (\xi_1, \xi_2) \in \mathcal{I}_1 \times \mathcal{I}_2, \qquad [1.19]$$

where the bivariate piecewise rational basis functions R_i^{pq} are computed from the bivariate B-spline functions N_i^{pq} as:

$$R_i^{pq}(\xi_1, \xi_2) = \frac{N_i^{pq}(\xi_1, \xi_2) w_i}{W(\xi_1, \xi_2)},$$

$$\text{with} \quad W(\xi_1, \xi_2) = \sum_{j=1}^{n_{2D}} N_j^{pq}(\xi_1, \xi_2) w_j. \qquad [1.20]$$

Therefore, the tensor product operation [1.18] is also involved for NURBS surfaces. As for NURBS curves, the weights give more shape control on the surface and allow us to describe exactly, for example, cylinders and spheres.

Let us eventually introduce the case of parametric volumes. These are trivariate functions obtained with an additional tensor product at the B-spline level. A NURBS volume can be expressed as follows:

$$V^h_{NS}(\xi_1, \xi_2, \xi_3) = \sum_{i=1}^{n_{3D}} R_i^{pqr}(\xi_1, \xi_2, \xi_3) P_i, \qquad [1.21]$$

where the trivariate piecewise rational basis functions R_i^{pqr} are given by:

$$R_i^{pqr}(\xi_1, \xi_2, \xi_3) = \frac{N_i^{pqr}(\xi_1, \xi_2, \xi_3) w_i}{W(\xi_1, \xi_2, \xi_3)},$$

$$\text{with} \quad W(\xi_1, \xi_2, \xi_3) = \sum_{j=1}^{n_{3D}} N_i^{pqr}(\xi_1, \xi_2, \xi_3) w_j,$$

$$\text{and} \quad N_i^{pqr}(\xi_1, \xi_2, \xi_3) = M_{i_1}^p(\xi_1) \times M_{i_2}^q(\xi_2) \times M_{i_3}^r(\xi_3). \qquad [1.22]$$

The trivariate B-spline basis functions above are the generalizations of those for B-spline surfaces (see equation [1.18]). Thus, a volume is equipped with three knot-vectors for each ξ_1-, ξ_2- and ξ_3-direction. Figure 1.10 depicts the case of a hollow cylinder (with similar geometry to that in Figure 1.2 (right)). Thanks to adapted weights, the circular part can be exactly represented. However, it has to be noted that only simple geometries can be modeled with a single spline patch. Indeed, given the tensor product structure at the B-spline level, a patch is necessarily a rectangle or a cuboid in the parameter domain for bivariate and trivariate topologies, respectively, which makes it poorly suited for representing complex, multiply connected domains. In this situation, non-conforming multipatch modeling that consists, roughly speaking, of assembling different (possible cut) spline patches is necessary. We do not go into further detail here. At this stage, we refer the reader to Cottrell *et al.* (2009) and indicate that the multipatch modeling issue is addressed, in detail, in this book.

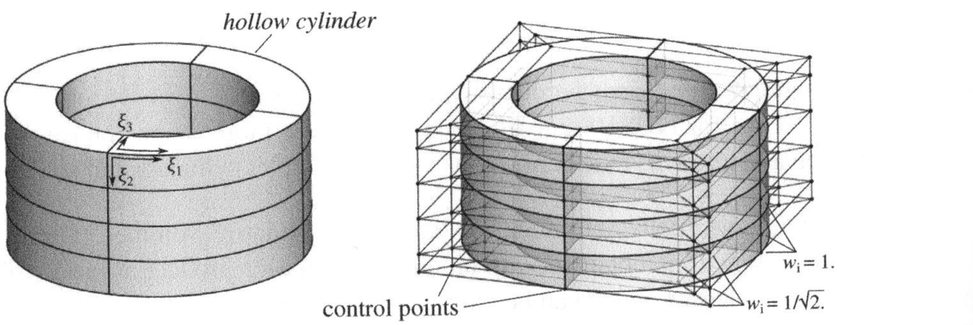

Figure 1.10. *Example of a hollow cylinder described with a quadratic NURBS volume*

In the rest of this book, we may omit the degree of the functions when unnecessary. In this case, the notations of the form R_i^{pqr} will simply turn into R_i. In addition, we will sometimes use a matrix form to define the spline objects for ease of reading, with respect to the implementation. In this respect, the general expression for NURBS (thus also including B-spline) geometries with parameter $\xi \in \mathcal{I}_d \subset \mathbb{R}^d$ (with d being the parametric dimension of the spline entity, i.e. $d = 1$ for curves, $d = 2$ for surfaces and $d = 3$ for volumes) will be simply written as:

$$G^h(\xi) = \sum_i R_i(\xi) P_i = \mathbf{R}^T(\xi)\mathbf{x}, \quad \forall \xi \in \mathcal{I}_d, \qquad [1.23]$$

where $\mathbf{R}$ and $\mathbf{x}$ denote the matrix of the NURBS basis functions and the vector collecting the locations of the associated control points, respectively. Considering, for example, control points in 3D, $\mathbf{R}$ and $\mathbf{x}$, formally read as:

$$\mathbf{R} = \begin{bmatrix} R_1 & 0 & 0 \\ R_2 & 0 & 0 \\ \vdots & \vdots & \vdots \\ 0 & R_1 & 0 \\ 0 & R_2 & 0 \\ \vdots & \vdots & \vdots \\ 0 & 0 & R_1 \\ 0 & 0 & R_2 \\ \vdots & \vdots & \vdots \end{bmatrix} \quad \text{and} \quad \mathbf{x} = \begin{bmatrix} x_1 \\ x_2 \\ \vdots \\ y_1 \\ y_2 \\ \vdots \\ z_1 \\ z_2 \\ \vdots \end{bmatrix}. \qquad [1.24]$$

P_i in [1.23] is obviously the subset of $\mathbf{x}$ gathering the coordinates (x_i, y_i, z_i) of the ith control point.

1.2.3. *Design features and shape parameterization*

Besides being attractive for generating geometries, B-splines and NURBS are interesting because they offer great flexibility in terms of shape control. This certainly explains why most of the CAD software is based on these spline technologies. In this work, which ultimately intends to perform the shape optimization of structures, we will deeply benefit from many of the appealing properties of B-splines and NURBS, especially their efficient refinement procedures and their flexible shape manipulation possibilities. Key aspects regarding these two features are given in the following.

1.2.3.1. *Refinement*

B-splines and NURBS can be refined while leaving the underlying geometry intact. There are different refinement mechanisms. The first possibility consists of inserting new knots into the knot-vectors. This process is called *knot-insertion*. More precisely, let us take the case of a p-degree B-spline curve, defined from knot-vector $\Xi_1 = \{\xi_1^1, \cdots, \xi_1^{n_1+p+1}\}$, with n_1 control points P_i. Knot-insertion consists of adding k_1 new (interior) knots $\hat{\xi}_1^1, \ldots, \hat{\xi}_1^{k_1}$ in the knot-vector. The new refined version of the B-spline curve takes the same degree p, but is built from the new knot-vector, which concatenates the initial knots ξ_1^i with the inserted ones $\hat{\xi}_1^j$ (in ascending order). The new knot-vector $\hat{\Xi}$ counts $n_1 + k_1 + p + 1$ knots. Ultimately, the only input that remains to be able to define the refined spline object is the vector collecting the control points. In other words, we need to compute the coordinates of the $n_1 + k_1$ control points denoted here $\hat{P}_i$. Interestingly, it takes the form of a linear application:

$$\hat{\mathbf{x}} = \mathbf{D}^p\mathbf{x} \qquad [1.25]$$

where $\mathbf{D}^p$ is the refinement matrix (size $l(k_1 + n_1) \times ln_1$, with l being the dimension of the physical space) and vectors $\hat{\mathbf{x}}$ and $\mathbf{x}$ gather the control point positions associated with $\hat{P}_i$ and P_i, respectively (the same notation as in [1.23]). The refinement matrix is sparse and is obtained recursively (Cottrell *et al.* 2009), since it is based on formula [1.13]:

$$\mathbf{D}_{ij}^0 = \begin{cases} 1 & \text{if } \xi_1^j \leq \hat{\xi}_1^i < \xi_1^{j+1} \\ 0 & \text{otherwise} \end{cases}, \qquad [1.26]$$

$$\mathbf{D}_{ij}^{q+1} = \frac{\hat{\xi}_1^{i+q} - \xi_1^j}{\xi_1^{j+q} - \xi_1^j}\mathbf{D}_{ij}^q + \frac{\xi_1^{j+q+1} - \hat{\xi}_1^{i+q}}{\xi_1^{j+q+1} - \xi_1^{j+1}}\mathbf{D}_{ij+1}^q, \; q = 0, 1, .., p-1. \qquad [1.27]$$

An example of refinement through knot-insertion, starting with a quadratic B-spline curve, is presented in Figure 1.11 (middle). We can observe the changes for the basis functions and for the control points, while leaving the initial geometry intact.

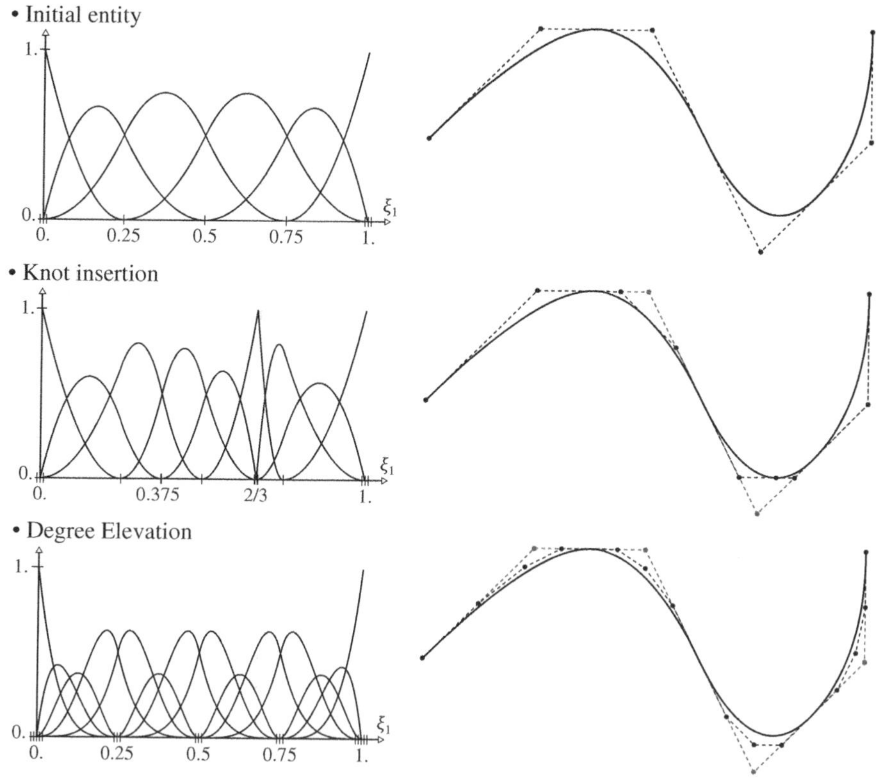

Figure 1.11. *Refinement procedures applied to a B-spline curve*

The other refinement possibility is the *degree elevation*. As its name suggests, it consists of increasing the polynomial degree of the B-spline basis functions. Hence, it has to be noted that through the order-elevation, all values in the knot-vector are repeated to preserve the initial discontinuities of the derivatives of the functions (ensuring that the refined spline space includes the initial coarse spline space). As an illustration, let us take the example from Figure 1.11. The initial knot-vector Ξ is for quadratic B-spline functions while the refined knot-vector $\hat{\Xi}$ is for quartic functions. More precisely, Ξ and $\hat{\Xi}$ read:

$$\begin{aligned} \Xi &= \{0, 0, 0,\ 0.25, 0.50, 0.75,\ 1, 1, 1\}, \\ \hat{\Xi} &= \{0, 0, 0, 0, 0, 0.25, 0.25, 0.25, 0.50, 0.50, 0.50, .. \\ &\quad .., 0.75, 0.75, 0.75, 1, 1, 1, 1, 1\}\,. \end{aligned} \qquad [1.28]$$

At the (interior) knot location, the regularity of the refined curve is still C^1:. Indeed, we have $p - m = 4 - 3 = 1$ (with m being the multiplicity of the interior knots). The computation of the new control points can be performed in three steps (Piegl and Tiller 1997):

1) subdivide the curve into as many (C^0) Bézier curves as there are elements by repeating p times the interior knots;

2) elevate the polynomial degree on each of these segments;

3) remove exceeding knots to combine the Bézier element curves into one B-spline curve.

Steps 1 and 3 are performed using the knot-insertion procedure. The formula for the degree-elevation of a Bézier curve can be found for the example in Piegl and Tiller (1997). It is possible to elevate the degree from p to $p + t$ in one step:

$$Q_i^t = \sum_{j=\max(0,i-t)}^{\min(p,i)} \frac{\binom{p}{j}\binom{t}{i-j}}{\binom{p+t}{i}} Q_j, \quad i = 0, \ldots, p+t, \qquad [1.29]$$

where Q_i and Q_i^t are the control points of the initial Bézier curve of degree p and of the new Bézier curve of degree $p + t$, respectively (the numbering of control points starts at zero, here). Overall, the interesting point to note is that the degree-elevation procedure also takes the form of a linear application. A refinement matrix can be built to express the new control points of the degree-elevated B-spline entity in relation to the initial ones (Lee and Park 2002):

$$\hat{\mathbf{x}} = \left[\mathbf{S}^r \mathbf{T}^t \mathbf{D}^i\right] \mathbf{x}, \qquad [1.30]$$

where $\mathbf{D}^i$ is the matrix that decomposes the curve in multiple Bézier curves, $\mathbf{T}^t$ is the matrix that elevates the degree of the Bézier curves (see equation [1.29]) and $\mathbf{S}^r$ is the matrix that recomposes the B-spline from the multiple Bézier curves. The matrix $\mathbf{S}^r$ can be seen as a least-squares pseudo-inverse of the operator used for knot-insertion in step 1.

The refinement procedures (knot-insertion and degree-elevation) can be combined, which offers great flexibility regarding the parameterization of a

B-spline curve. In particular, it enables us to introduce a new refinement scheme in which the polynomial degree and the regularity of the basis functions can be simultaneously increased. In this case, degree-elevation is performed first, and then the new knot values are inserted (with multiplicity one), so that elements are added while ensuring the maximum available regularity of the basis functions at the knots level, namely C^{p-1}. This refinement procedure is called k-refinement (Cottrell *et al.* 2007). In this work, unless otherwise stated, we will perform k-refinement to take advantage of the superior properties of splines. Eventually, we can simply denote by $\mathbf{D}$ the refinement matrix, independent of the refinement procedure. Let us also mention that the same refinement operators can be used for NURBS curves by using the homogeneous coordinates, i.e.:

$$\hat{\mathbf{x}}^w = \mathbf{D}\mathbf{x}^w, \qquad [1.31]$$

where $\mathbf{x}^w$ has one more component than $\mathbf{x}$ for each control point: considering $P_i = (x_i, y_i, z_i)$ with associated weight w_i, $P_i^w = (x_i w_i, y_i w_i, z_i w_i, w_i)$. The refinement strategy can also be extended to surfaces and volumes: the refinement is performed direction per direction in these situations. For more details on the refinement strategies of splines and their matrix representations, the reader is referred to Piegl and Tiller (1997), Lee and Park (2002) and Cottrell *et al.* (2007).

NOTE.– If the B-spline and NURBS technologies offer new opportunities with the k-refinement process, it has to be noted that multivariate B-spline or NURBS bases do not provide a natural possibility for local mesh refinement. Again, the reason for this is the rigid tensor-product structure [1.18] of these functions that precludes local mesh refinement within a patch. The possible harmful effects of a tensor product structure have already been underlined in Figure 1.8 (bottom). Basically, two approaches exist to answer this issue: (i) the generation and coupling of (usually non-conforming) spline patches and (ii) the creation of alternative splines that enable local mesh refinement. In this book, and particularly in Chapter 2, we will contribute to approach (i) in order to perform local mesh refinement and incorporate more general local phenomena within a B-spline or NURBS model. Regarding approach (ii), we may cite, for example, the hierarchical B-splines and NURBS (Vuong *et al.* 2011; Scott *et al.* 2014; Hennig *et al.* 2016), the locally refined B-splines (Dokken *et al.* 2013), the T-splines (Bazilevs *et al.* 2010; Scott *et al.* 2012; Veiga *et al.* 2012) and the hierarchical T-splines (Evans *et al.* 2015; Chen and

Borst 2018). Among these strategies, T-splines have gathered an important momentum since they are also suitable for addressing trimmed multipatch geometries (see the discussion in section 1.4.1). Nevertheless, if we seek to perform local mesh refinement only, the hierarchical splines may be easier to implement (D'Angella *et al.* 2018). Finally, note that the work on new splines is by no means complete, as demonstrated, for example, by the recent introduction of U-splines (unstructured splines) by Coreform for industrial applications (Thomas *et al.* 2018).

1.2.3.2. *Geometric modification*

As stated previously, B-splines and NURBS give the designer great flexibility over the shape. We have already observed the influence of the knot values (see Figure 1.4) and the influence of the weights (see Figure 1.7). The designer can also operate smooth shape changes by modifying the coordinates of the control points, as shown in Figure 1.12. This possibility will be of great interest when performing shape optimization. The design variables could be the spatial location of the control points of the spline entities, which, given the high degree of regularity of spline functions, will enable us to get a light and smooth parameterization of the geometry and of its update. This point will be discussed further in Chapter 4.

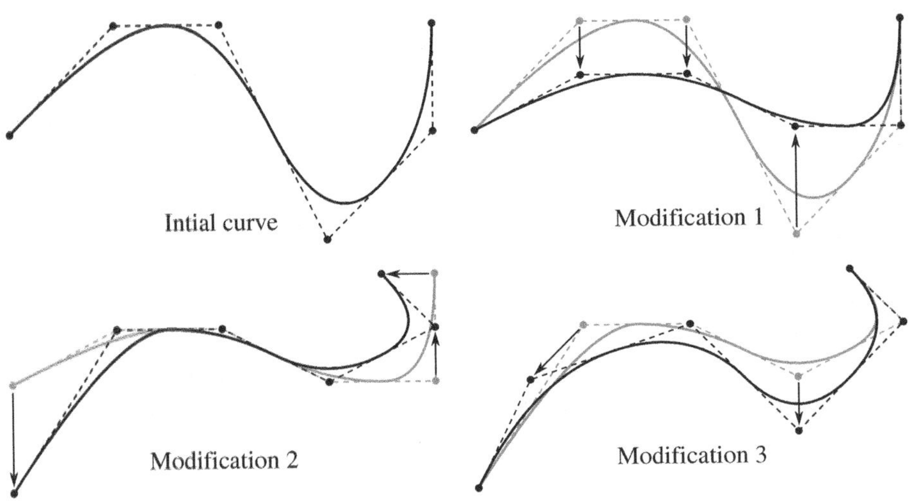

Figure 1.12. *Example of shape modification by altering the spatial location of the control points*

1.2.4. *Spline-based finite element analysis: isogeometric principle*

1.2.4.1. *General idea*

The main idea behind IGA is to use the spline-based parameterizations discussed above to build the approximation subspaces when applying Galerkin's method for solving partial differential equations (Hughes *et al.* 2005; Cottrell *et al.* 2009). In this framework, the method can be viewed as a generalization of the FEM that considers smooth and higher-degree functions to replace typical Lagrange polynomials in the computations. Within this work, we recall that only NURBS (which constitute the most commonly used technology in CAD) and simpler B-splines are considered. In the end, this is the only major difference with the classical FEM, from an analytical point of view. IGA relies on the same variational formulations as standard FEM. When invoking the isoparametric element concept, the spline basis turns these weak forms into systems of algebraic equations. In solid mechanics, it means that the displacement field is approximated as:

$$u^h(\xi) = \sum_i R_i(\xi) u_i, \qquad [1.32]$$

where R_i are the spline basis functions originated from the considered geometric model of the structure (see equation [1.23]). With this numerical approximation in hand, the principle of virtual work gives the following well-known linear system:

$$\mathbf{Ku} = \mathbf{f}, \qquad [1.33]$$

where $\mathbf{K}$ is the so-called stiffness matrix, $\mathbf{u}$ is the displacement vector (vector collecting the degrees of freedom (DOF)) and $\mathbf{f}$ is the force vector. We do not provide more information here on the expressions of these operators. In the rest of this chapter, these quantities will either be the very classic ones (context of linear elasticity, for example), or detailed in more specific situations (e.g. context of thin shell analysis).

1.2.4.2. *Performance*

Beyond the reinforced link between CAD and numerical simulation, IGA turned out to be a superior scientific computing technology. It is now often seen as a high-performance computing method. The reason for this is the higher regularity of spline-based functions that enables us to capture physical solutions with fewer DOF than in standard FEM (Evans *et al.* 2009). In this respect, IGA has now been successfully applied to numerous disciplines of computational mechanics, such as shell analysis (Kiendl *et al.* 2009; Echter *et al.* 2013; Bouclier *et al.* 2015b), contact problems (Seitz *et al.* 2016; Antolin *et al.* 2018), fluid–structure interaction (Kamensky *et al.* 2017; Apostolatos *et al.* 2019), biomechanics (Vilanova *et al.* 2013; Morganti *et al.* 2015; Patelli *et al.* 2017), geometrical and material nonlinear solid mechanics (Elguedj *et al.* 2008; Lipton *et al.* 2010; Bouclier *et al.* 2015a; Ambati *et al.* 2018), fluid dynamics (Bazilevs *et al.* 2007; Akkerman *et al.* 2011) and structural vibration (Cottrell *et al.* 2006; Shojaee *et al.* 2012), to name a few. In this book, the idea is to benefit from the capabilities of IGA and to arrive at the shape optimization of complex structures, which undoubtedly constitutes one of the most valuable applications for IGA, as both its superior analysis and attractive geometrical properties can magnificently express.

NOTE.– When dealing with the performance of IGA, another important aspect to mention concerns the numerical integration used to evaluate the stiffness operator $\mathbf{K}$ (see equation [1.33]). In this work, we proceed, as originally, in IGA, which is the same as in the FEM standard: we perform a classic looping over the knot-span elements, and we take $p+1$ Gauss points per element and per parametric direction (with p being the polynomial degree of the shape functions in the corresponding direction). However, we quote that this procedure is not optimal given the higher regularity of the spline functions. Numerous works that aimed to reduce the cost of the construction of isogeometric operators have emerged since the advent of the concept. In this respect, we may cite the development of advanced quadrature rules, more or less related to the entire isogeometric patch (Hughes *et al.* 2010; Auricchio *et al.* 2012a; Bouclier *et al.* 2012; Schillinger *et al.* 2014; Adam *et al.* 2015). In the same idea, it may be preferred to focus on collocated-IGA (Auricchio *et al.* 2010, 2012b) when the polynomial degree is very high. Finally, it was recently noted in the standard Galerkin framework that a revisit of the standard looping over elements in the assembly process could allow IGA to meet its full efficiency (Calabro *et al.* 2017; Sangalli and Tani 2018; Hiemstra *et al.* 2019). These innovative implementation techniques are not considered in this work.

1.3. Improved CAD-CAE integration for robust optimization

With the isogeometric concept and the underlying spline technologies in hand, let us now step back and briefly retrace the original motivation behind IGA, i.e. the opportunity to reunify computer-aided design (CAD) and computer-aided engineering (CAE). Pushing forward the reasoning also enables us to envisage the power of the approach for shape optimization, as discussed at the end of this section.

1.3.1. *Returning to the original motivations behind IGA*

As previously mentioned, IGA was originally introduced by Hughes *et al.* (2005) with a principal motivation: to reunify the worlds of CAD and CAE, which, in past decades have evolved independently (although they are closely related during engineering design and certification). Engineering design involves, on one side, the designer who generates geometric models through CAD files, and on the other side, the stress analyst who builds the inputs according to a numerical code (that calls upon FEM most of the time). The problem is that the translation of the CAD file into the FEM inputs is not straightforward in practice: an important analysis preparation step is necessary, as illustrated in Figure 1.13(a).

Note that the issue to move from design and analysis does not only concern the mesh generation, but also the efficient creation of appropriate, *analysis-suitable* geometries (Cottrell *et al.* 2009). In fact, the designer endeavors to represent, as faithfully as possible, a conceptual geometry. It means that the generated CAD model usually presents many details, for example holes, fillets and numerous sub-parts, such as bearings. Highly detailed CAD models are rarely exploitable for the simulation since it usually contains details that may be viewed as irrelevant by the analyst. As a result, in addition to the mesh generation step, a lot of time is spent generating analysis-suitable geometries, where the initial CAD model is *defeatured*. For sophisticated engineering systems, things can turn out to be very tricky. Different simulations such as, for example, structural mechanics, fluid dynamics, heat transfer and many others are performed. Each type of analysis has its own particular method, and thus different models are required for each of them. Figure 1.14 illustrates this point: for sophisticated engineering systems, for example, here an airplane, different visions of the same product may exist depending on the physical field. It follows that appropriate *simulation-specific* geometries are needed. Still looking at the example of Figure 1.14, the aerodynamicist may focus on the outer geometry of the aircraft, whereas the stress analyst is mainly concerned by the internal structure. Therefore, there is not a unique CAD model of the airplane. A lot of geometric models are required, depending on the simulation to be performed.

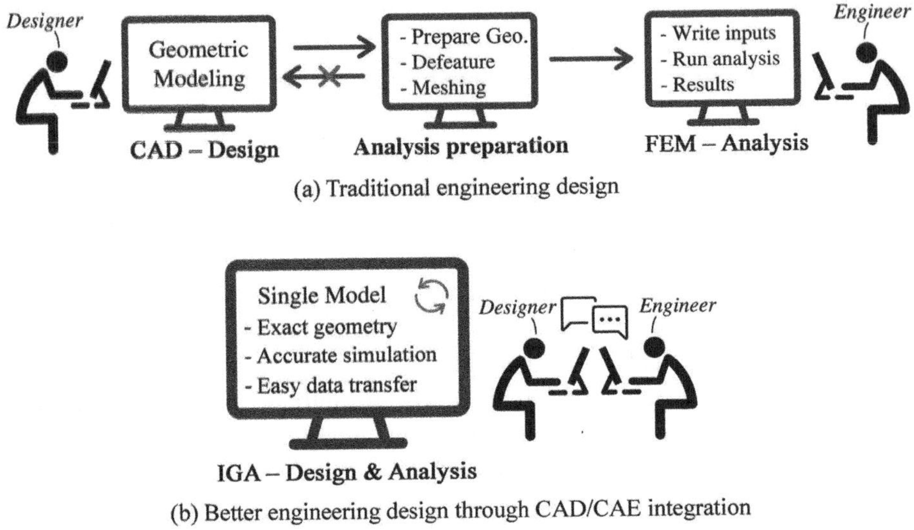

Figure 1.13. *The original motivation of IGA is to improve engineering design and certification by integrating Design and Analysis into a single model: henceforth, designers and engineers can collaborate more intimately. For a color version of this figure, see www.iste.co.uk/bouclier/IGA.zip*

Figure 1.14. *Engineering systems: each engineer has their own vision and requirements for the same product (adapted from artwork by C.W. Miller)*

According to Cottrell *et al.* (2009), the translation to analysis-suitable geometries, in addition to the meshing and the input setting for simulation codes, take *de facto* over 80% of the overall analysis time. It is thus clear that a stronger dialog between CAD and CAE would be of significant benefit. The engineer should be aware of what the designer does during the whole design and certification process, and vice versa. IGA brings this issue back to the table and has fostered fruitful discussions regarding a better CAD-CAE integration. As explained in Figure 1.13(b), the goal of IGA can be seen as an attempt to ease the collaboration between the designers and the engineers by integrating efficient geometric modeling possibilities and accurate simulation capabilities into a single model. As of now, it is nevertheless important to state that the dream process of Figure 1.13(b) is still not truly possible in practice, even with IGA. Indeed, we will see in section 1.4 that the prerequisites for geometric modeling and numerical simulations are too different today to envisage a direct link between design and analysis without any additional effort from the two communities. Things are starting to move forward thanks to the arrival of IGA in some commercial FE packages (see, for example, Hartmann *et al.* (2011, 2016); Duval *et al.* (2015); Lai *et al.* (2017) for implementations in LS-Dyna, Abaqus and Radioss, respectively). But, now more than ever, both disciplines need to work together in order to handle this common goal and define the future of engineering design. Contributions, from an analytical point of view, are present in this book, to help bridge the remaining gap between design and analysis in the context of IGA.

1.3.2. *An ideal framework for parametric shape optimization*

Although a proper introduction of structural isogeometric shape optimization will not be given until Chapter 4 (see, in particular, section 4.2 for a general discussion), we can already suspect the benefit of combining high-quality geometric representations and efficient analysis capabilities into one single model for this area of research. Indeed, without providing details yet, shape optimization involves an iterative process with three principal steps, as explained in, for example, Bletzinger (2014) and illustrated in Figure 1.15. Depending on the choices performed to describe the design model (i.e. the geometrical model where the design variables allowing for shape modification are defined; see the top box in Figure 1.15), the transitions between these steps (the blue arrows in Figure 1.15) can become quite tedious in classic FEM.

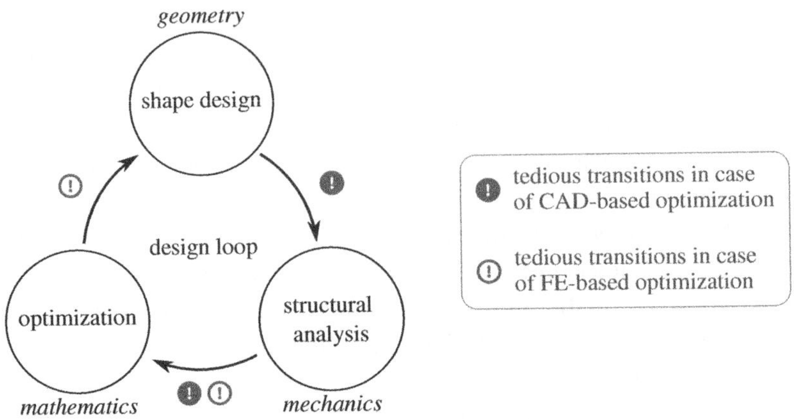

Figure 1.15. *Main steps for shape optimization (inspired from Bletzinger (2014)): difficulties can occur during the transition from one step to the other when either the CAD-based or the FE-based approach is used. For a color version of this figure, see www.iste.co.uk/bouclier/IGA.zip*

Basically, two main approaches exist in standard FEM to carry out the shape optimization of structures. The first one, usually classified as the *FE-based optimization* alternative in the field, consists of using the spatial location of the FE nodes as the design variables (i.e. the variables that update the geometrical shape) (Bletzinger *et al.* 2010; Le *et al.* 2011; Firl *et al.* 2013; Bletzinger 2014). The interest is that the design model is directly usable for structural analysis (see the bottom-right box in Figure 1.15). However, important difficulties are encountered when treating the large number of design parameters. Indeed, the number of nodes of an FE mesh is often significant. Thus, delicate mesh regularization techniques need to be put in place to prevent the appearance of irregular, "unrealistic" shapes (Firl *et al.* 2013; Bletzinger 2014) (see top-left arrow in Figure 1.15). In the same idea, special care is needed to exploit the results coming from the FE structural analysis, such that appropriate shape updates could be computed through the optimization process (see the bottom arrow in Figure 1.15). Furthermore, at the end of the process, it is almost always required to convert the FE results in terms of CAD-suitable data for production. Conversely, the second class of techniques, denoted by the *CAD-based approach*, proposes to resort to the same spline-based functions as in CAD software to describe the design model (Braibant and Fleury 1984; Olhoff *et al.* 1991; Ramm *et al.* 1993). Hence, this time, the design variables are the spatial locations of the control points of the spline entities, which bring very appealing features to proceed the shape design step (see discussion in section 1.2.3). The difficulty with this second family of methods is then transferred to the connection of the design model and the analysis model (Hasan Imam 1982; Bletzinger *et al.* 1992) (see the top-right arrow in Figure 1.15). Indeed, in the CAD-based approach, the design model is based on splines, while the analysis model is based on standard FE. A meshing step separates both models. This also creates drawbacks when exploiting the results from the structural analysis (see the bottom arrow in Figure 1.15) since these come from the FE analysis model and should be transferred back to the spline design model.

Benefiting from the discussion of section 1.3.1, it is obvious that one way to naturally communicate between the design and analysis models is to use the IGA framework, as it enables us to get an analysis model made of splines functions. Hence, isogeometric shape optimization has been successfully applied to a wide range of applications since the advent of IGA (see Wall *et al.* (2008); Nagy *et al.* (2010, 2011, 2013); Qian (2010); Kiendl *et al.* (2014); Taheri and Hassani (2014); Fußeder *et al.* (2015); Wang and Turteltaub (2015); Kang and Youn (2016); Herrema *et al.* (2017); Lian *et al.* (2017); Choi and Cho (2018); Lei *et al.* (2018); Hirschler *et al.* (2019b); Weeger *et al.* (2019) among others). It concerns not only structural shape optimization, but also other fields, such as heat conduction (Wang *et al.* 2017b), electromagnetics (Nguyen *et al.* 2012; Dang Manh *et al.* 2014), fluid mechanics (Park *et al.* 2013) and many other optimization problems.

1.4. The analysis-suitable model issue

Although based on the same basis functions, the link between CAD and analysis, originally advocated by IGA, is not trivial in practice. Indeed, standard IGA requires a boundary-fitted discretization for the analysis, while in CAD programs, where the only matter is the rendering of the geometry, entities are described as a collection of their boundary surfaces that can be joined and/or intersected through Boolean operations. When the artificial parts of the surfaces are cut away, the visualization of the resulting object changes, while its mathematical description does not. To clarify this point, we need to introduce the concept of trimming and the related issue of non-conforming multipatch analysis. These two challenges are currently the object of intensive studies from both the geometric modeling and scientific computing communities, since they are primordial to truly bridge the gap between CAD and analysis.

1.4.1. *The trimming concept*

Trimming is one of the most fundamental tools in CAD environments, where complex geometries are built using Boolean operations. The process enables us to create an almost unlimited range of geometric shapes. The trimming concept is illustrated in 2D for the simple situation of a circular hole living in a rectangular structure (see Figure 1.16). This surface can be classified as a trimmed surface. Its description is simply given by: a one-patch (linear, 1 element) B-spline surface parameterization for the plate (without the hole) and a (quadratic, four elements) NURBS curve parameterization for the trimming curve that forms the boundary of the hole (see Figure 1.16(a)). In CAD, the trimming curve specifies visible and invisible regions on the surface patch (see Figure 1.16(b)). As a result, the underlying spline patch remains unaffected by the trimming object and preserves its topology. Conversely, using standard IGA for the analysis of such a geometry would require a delicate *spline re-parameterization* of the whole model, leading to the appearance of C^0 lines due to the patch structure of multivariate splines. An example of a boundary-fitted NURBS parameterization of the plate including the hole (i.e. without trimming) is shown in Figure 1.16(c). It is composed of four elements connecting along C^0 lines with degree 2 into the circumferential direction and degree 1 into the radial direction. In the case of multiple holes, the situation becomes even trickier since the resulting geometry strongly differs topologically from a square. As a result, it becomes necessary to define different patches around each hole (see Figure 1.16(d) for an illustration with two holes) and to group them together (the same process as assembling finite elements in standard FEM). The new spline models of Figure 1.16(c) and 1.16(d) are commonly referred to as *analysis-suitable* models in the sense that they can be easily enhanced using classic spline refinement (see section 1.2.3) to compute the solution of a corresponding mechanical problem.

The simple example of a plate with holes in Figure 1.16 underlines the difficulties of generating analysis-suitable models for IGA. From a more general point of view, there are mostly two ways to deal with this issue. First, we can strive to remove all trimmed regions by invoking re-parameterization strategies (as in Figure 1.16(c) and 1.16(d)), which can be cumbersome (Xu *et al.* 2013; Akhras *et al.* 2016; Massarwi *et al.* 2019). The second option consists of developing advanced numerical tools for analyzing models with the mentioned defects coming from the geometric modeling (Marussig and Hughes 2018; Teschemacher *et al.* 2018). The works presented in this book are heading in the second direction: in particular, trimming will be faced in Chapter 2 for the simulation of local phenomena within IGA models. Let us finally quote that the two approaches are not opposed, and, as already stated, it is by combining the innovations from both the geometrical and analysis sides that researchers and engineers will succeed in the long term.

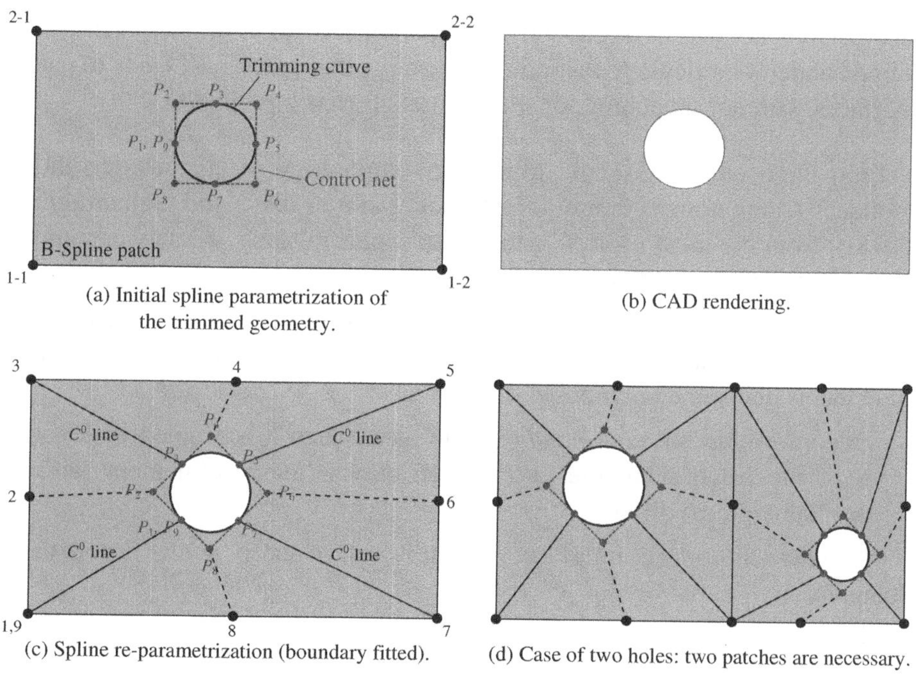

(a) Initial spline parametrization of the trimmed geometry.

(b) CAD rendering.

(c) Spline re-parametrization (boundary fitted).

(d) Case of two holes: two patches are necessary.

Figure 1.16. *Illustration of the trimming concept on a simple example of a plate with holes. For a color version of this figure, see www.iste.co.uk/bouclier/IGA.zip*

1.4.2. *Non-conforming multipatch parameterization*

Unfortunately, trimming, as previously introduced, is not the only challenge in practice. A related question is the handling of complex non-conforming multipatch models. Indeed, a raw geometric model, as given by a CAD modeler, usually involves plenty of trimmed patches that arbitrarily intersect with each other. More specifically, again because of the rigid tensor product structure of (standard) multivariate spline basis, a single B-spline or NURBS patch cannot represent complex industrial structures. Instead, multiple (possibly trimmed) spline patches are required, where each of them describes a specific part of the overall structure. Multipatch modeling is thus inevitable when dealing with structures with non-trivial geometries using IGA. The main issue from an analytical point of view concerns the treatment of the junctions between those patches. Figure 1.17 illustrates the problem in relatively simple situations. The difficulty lies in the fact that geometric modeling does not pay special attention to the presence of multiple patches: each patch is defined independently. The parameterization of one patch is not defined under a particular constraint due to any other patches. It leads to patch interfaces with non-matching, or even non-conforming properties.

From here on, we use the following terminology to characterize those interfaces (same nomenclature as in Bouclier *et al.* (2017) and Guinard *et al.* (2018)); from the most restrictive to the most general case, we may encounter the coupling situation across:

– a *matching* interface, when it is aligned with some edges of the patches and the discretizations on both sides are exactly the same (same mesh refinements and same polynomial degrees);

– a conforming but *non-matching* interface, when it is aligned with some edges of the patches but the two discretizations (mesh refinements and/or polynomial degrees) differ;

– a *non-conforming* interface, when it can overlap some (knot-span) elements.

Spline-based CAD models seldom, if ever, have matching interfaces. Depending on the geometry, isogeometric modeling can lead to non-matching interfaces, as depicted in Figure 1.17 (left). But in the more general case, non-conforming interfaces between the patches are to be expected (see Figure 1.17, right).

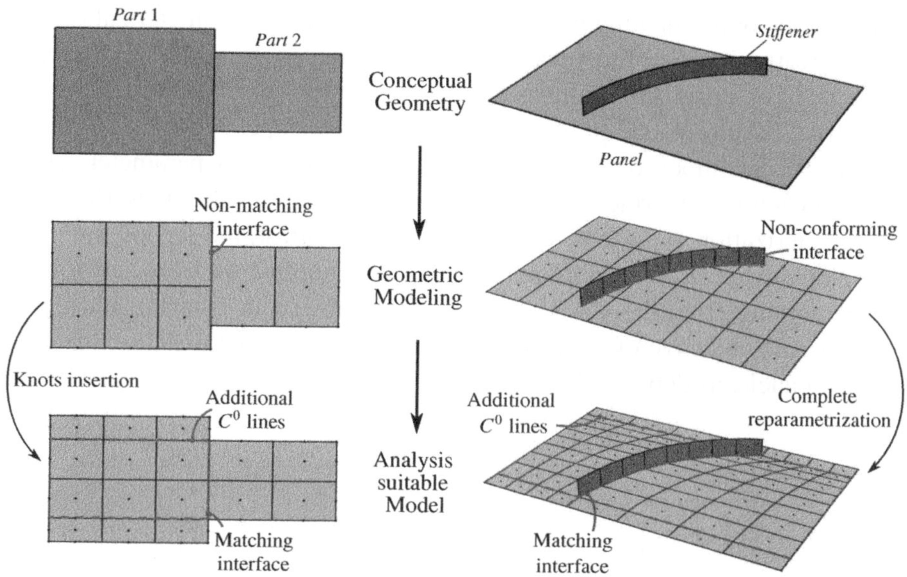

Figure 1.17. *B-spline and NURBS modeling generate multipatch geometries where each patch represents a specific part of the complete structure. Naturally, these patches undergo non-matching or even non-conforming parameterizations at their interfaces. Additional cumbersome procedures are required to get a final (standard) analysis-suitable model with conforming and matching interfaces. For a color version of this figure, see www.iste.co.uk/bouclier/IGA.zip*

For the analysis of multipatch models, the same two complementary approaches used for trimming can be listed. The first one calls upon *spline re-parameterization* procedures to transform the patch junctions into fully matching interfaces (Maquart 2019), so that the only work, from the analysis point of view, is to concatenate the duplicated control points at the interfaces. Once again, this task is far from trivial, from a geometrical point of view, in practice. In the simplest case of a non-matching but conforming interface (see Figure 1.17, left), knot-insertion (or removal) and degree-elevation (or reduction) make it possible to comply to the requirements for obtaining a matching interface. An illustration of the process for the situation in Figure 1.17 (left) is given in Figure 1.18. However, transforming non-conforming interfaces into fully matching interfaces is much more delicate (see Figure 1.17 (right) and Figure 1.19, which shows the trace spaces along the interface for this example). It cannot be done by invoking the fundamental geometric algorithms (knot-insertion, degree-elevation, etc.), as done in the non-matching case. Instead, a complete re-parameterization has to be performed since the local changes introduced to reach a matching interface have a global influence on the whole discretization. As an illustration, despite its simple in-plane rectangular shape, Figure 1.17 (right) needs to be decomposed into six patches to get an overall parameterization with a matching interface. The problem stems from the fact that this re-parameterization can hardly be automated. In contrast, the second option is to develop *advanced computational methods* for coupling the non-conforming patches coming from the CAD. This path is followed in Chapter 3 with, in particular, the objective of simulating real aeronautical structures (i.e. made of stiffened panels, as depicted in Figure 1.17, right).

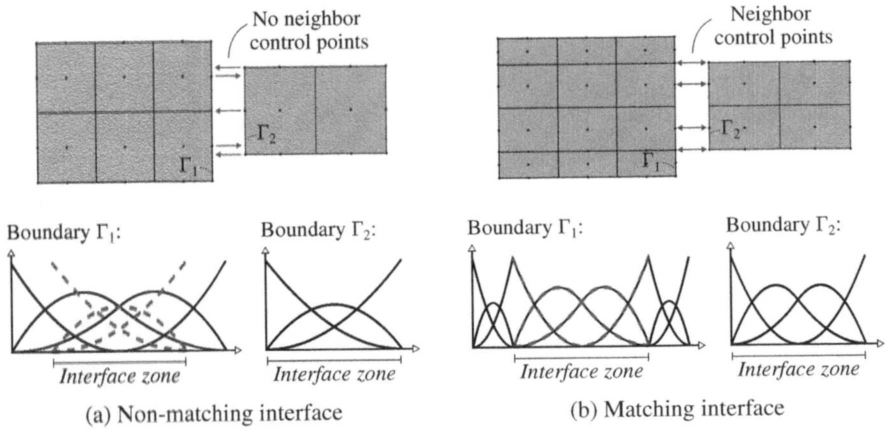

Figure 1.18. *Connecting patches that are initially non-matching (a): the basis functions of both patches need to be modified in order to become identical along the interface (b). For a color version of this figure, see www.iste.co.uk/bouclier/IGA.zip*

NOTE.– To truly bridge the gap between CAD and analysis in the case of arbitrary volumes, there is an additional challenge to overcome. Indeed, a volume is built in CAD through its boundary representation (B-rep) (Piegl 2005; Stroud 2006; Marussig and Hughes 2018), i.e. by grouping together intersected trimmed surface patches and filling the interior of the volume in the CAD rendering. Creating a solid analysis-suitable model or simulating solid structures directly from this CAD data also constitutes an active research topic in the field (Akhras *et al.* 2016; Massarwi and Elber 2016; Xu *et al.* 2017; Buffa *et al.* 2019; Massarwi *et al.* 2019). In this book, this issue is not addressed: we focus on either complex shell structures (thus the surface representation of the entity is directly available in CAD) or simple solid geometries.

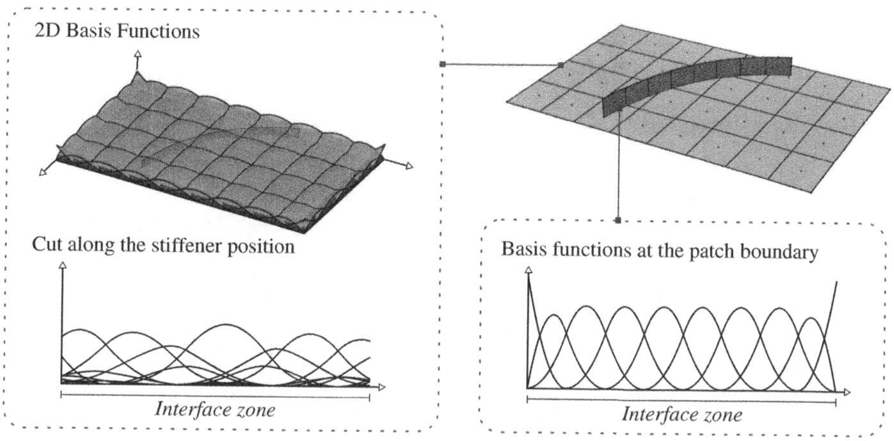

Figure 1.19. *Example of a non-conforming multipatch parameterization: the interface crosses the elements of the panel. The trace spaces along the interface of both sides are completely different. For a color version of this figure, see www.iste.co.uk/bouclier/IGA.zip*

1.4.3. *Imposing shape variation*

Developing advanced numerical schemes that are able to directly handle the trimmed and multipatch parameterizations of CAD models is all the more relevant, as the ultimate goal in this book is to perform the shape optimization of realistic structures. Indeed, we want to avoid modifying the parameterization of the whole complex model at every iteration of an optimization process. Illustrations are provided in Figures 1.20 and 1.21 for the case of the movement of a hole within a plate and of a stiffener on a panel, respectively. We seek to leave the discretization of the plate (respectively, panel) unchanged if only the position and shape of the hole (respectively, stiffener) are to be optimized. Thanks to trimming objects and non-conforming interfaces, each geometric sub-part has an independent parameterization, which facilitates the definition of its movements and its relations with a "good" analysis model (note that in the case of a plate with a hole, severe mesh distortion can occur with the boundary fitted alternative). Otherwise, this would lead to a method that is too cumbersome to be able to perform shape optimization properly, and in a reasonable time. Since it is of crucial importance for the computation and optimal design of complex structures, the discussion regarding the analysis-suitable model issue is not restricted to this section: additional insights will be given in the remaining chapters of this book.

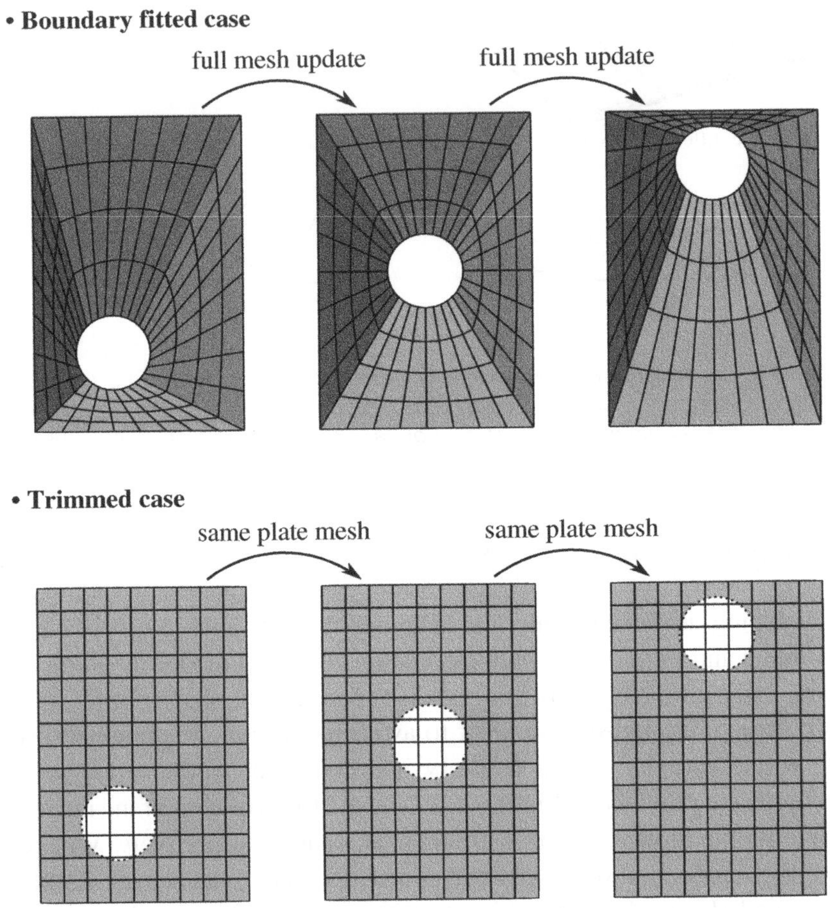

Figure 1.20. *The discretization of the plate needs to be updated at each modification of the hole when a boundary fitted model is used and mesh distortion may happen. Envisaging the trimmed model coming from CAD (i.e. the plate without the hole and the boundary of the hole) enables us to keep the mesh of the plate untouched, regardless of the position and shape of the hole. For a color version of this figure, see www.iste.co.uk/bouclier/IGA.zip*

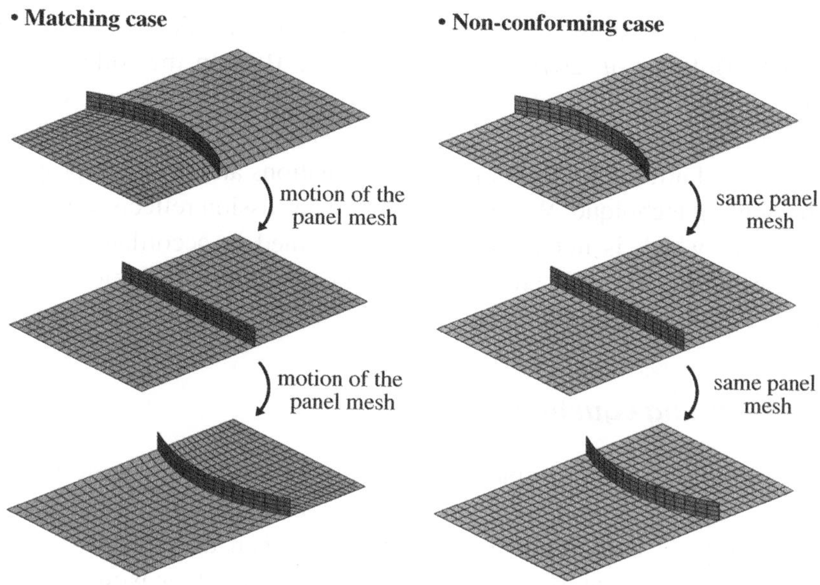

Figure 1.21. *The discretization of the panel needs to be updated at each shape modification of the stiffener when a matching multipatch model is used. Envisaging a model with non-conforming discretization enables us to keep the mesh of the panel untouched, regardless of the position and the shape of the stiffener. For a color version of this figure, see www.iste.co.uk/bouclier/IGA.zip*

1.5. Computation of non-conforming interfaces: a brief overview of usual weak coupling methods

As observed in the previous section, the development of advanced coupling methods that enable us to connect (possibly non-conforming) patches is crucial in IGA to simulate realistic structures and to answer our long-term objective of optimally designing such complex structures. This is also necessary in the search of a true bridge between CAD and analysis. To meet this goal, important studies have been performed in the last five years by the IGA community to revisit and extend the different weak domain coupling approaches that were initially developed for standard FEM. Basically, the resulting strategies can be divided into three classes: penalty coupling (Apostolatos *et al.* 2014a, 2014b; Herrema *et al.* 2019; Coradello *et al.* 2020; Leonetti *et al.* 2020), Mortar coupling (Brivadis *et al.* 2015; Dornisch *et al.* 2015, 2017; Bouclier *et al.* 2016, 2017; Coox *et al.* 2017a; Zou *et al.* 2018; Dittmann *et al.* 2019; Wunderlich *et al.* 2019) and Nitsche coupling (Nguyen *et al.* 2014; Ruess *et al.* 2014;

Jiang *et al.* 2015; Schillinger *et al.* 2016a; Bouclier and Passieux 2018; Hu *et al.* 2018; Buffa *et al.* 2019; Elfverson *et al.* 2019). In the following, we attempt to present the principle for each approach. In order to do so, we place ourselves in the context of linear elasticity and stay general by only giving the minimal state of art, the basic continuum formulations and the main properties of each coupling technique. We indicate that the discussion reflects the author's interpretation, which is not unique: it is performed in accordance with the works achieved in the next chapters, in particular the development of non-invasive and high-performance computing tools.

1.5.1. *Governing equations*

We consider here, the simple case of a bi-domain linear elastic problem in $\Omega \subset \mathbb{R}^d$, $d = 2$ or 3 being the parametric dimension of the domain. More precisely, Ω is divided into two disjoint, open and bounded subsets Ω_1 and Ω_2, joining at interface Γ (see Figure 1.22). In other words, we have $\Omega = \Omega_1 \cup \Omega_2 \cup \Gamma$ and $\Omega_1 \cap \Omega_2 = \varnothing$. We assume that the two non-overlapping sub-domains Ω_1 and Ω_2 are subjected to body forces f_1^g and f_2^g, respectively. Furthermore, forces F_1^g and F_2^g are associated with boundaries Γ_{F_1} and Γ_{F_2}, and displacements u_1^g and u_2^g are prescribed over boundaries Γ_{u_1} and Γ_{u_2}, respectively. The boundaries satisfy the following relations for $\mathrm{m} \in \{1, 2\}$ (see Figure 1.22):

$$\begin{aligned} &\Gamma_{F_\mathrm{m}} \cup \Gamma_{u_\mathrm{m}} \cup \Gamma = \partial\Omega_\mathrm{m} \,; \\ &\Gamma_{F_\mathrm{m}} \cap \Gamma_{u_\mathrm{m}} = \varnothing \,; \\ &\Gamma_{F_\mathrm{m}} \cap \Gamma \quad = \varnothing \,; \\ &\Gamma_{u_\mathrm{m}} \cap \Gamma \quad = \varnothing . \end{aligned} \qquad [1.34]$$

In each sub-domain, the kinematic constraints, the equilibrium equations and the constitutive relations have to be verified. Using again, the subscript m to denote a quantity that is valid over region Ω_m (with $\mathrm{m} \in \{1, 2\}$), the corresponding governing equations read:

$$\mathrm{div}(\sigma_\mathrm{m}) + f_\mathrm{m}^g = 0 \qquad \text{in} \quad \Omega_\mathrm{m} \,; \qquad [1.35a]$$

$$\sigma_\mathrm{m} = C_\mathrm{m}\, \varepsilon\,(u_\mathrm{m}) \qquad \text{in} \quad \Omega_\mathrm{m} \,; \qquad [1.35b]$$

$$u_\mathrm{m} = u_\mathrm{m}^g \qquad \text{over} \quad \Gamma_{u_\mathrm{m}} \,; \qquad [1.35c]$$

$$\sigma_\mathrm{m}\, n_\mathrm{m} = F_\mathrm{m}^g \qquad \text{over} \quad \Gamma_{F_\mathrm{m}} \,. \qquad [1.35d]$$

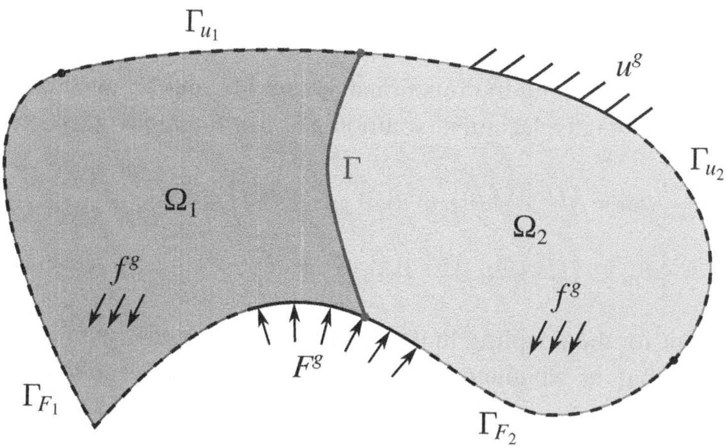

Figure 1.22. *Reference domain coupling problem. For a color version of this figure, see www.iste.co.uk/bouclier/IGA.zip*

In the above equations, $\varepsilon\left(u_{\mathrm{m}}\right)$ denotes the infinitesimal strain tensors, σ_{m} denotes the Cauchy stress tensors, C_{m} denotes the Hooke tensors and n_{m} represents the outward unit normal to Ω_{m}. We specify here that we perform, as follows, the notations in the document: continuous quantities are in normal type, while discrete quantities (i.e. vectors and matrices) are in boldface type (see equation [1.33]). To complete the formulation of the boundary value problem, the interface condition has to be added:

$$u_1 - u_2 = 0 \quad \text{over } \Gamma\,; \qquad [1.36a]$$

$$\sigma_1 n_1 + \sigma_2 n_2 = 0 \quad \text{over } \Gamma. \qquad [1.36b]$$

It ensures the kinematic compatibility and equilibrium of the tractions, respectively, along the coupling interface Γ between the two sub-domains. Combining conditions [1.36] with local equations [1.35] is strictly equivalent to a linear elasticity problem posed over the whole initial domain Ω. In the case of arbitrary non-matching and non-conforming interfaces, the coupling conditions cannot be imposed in a strong way. Instead, those coupling conditions are prescribed in the weak sense, and thus, the coupling methods differ by the way those weak conditions are formulated.

1.5.2. *Penalty coupling*

Let us start by defining the functional spaces $\mathcal{U}_{\mathrm{m}}$ and $\mathcal{V}_{\mathrm{m}}$ over domain Ω_{m} that will contain the displacement solution and test functions, respectively:

$$\mathcal{U}_{\mathrm{m}} = \left\{ u_{\mathrm{m}} \in \left[H^1\left(\Omega_{\mathrm{m}}\right)\right]^d,\ u_{\mathrm{m}}|_{\Gamma_{u\mathrm{m}}} = u_{\mathrm{m}}^g \right\} \ ;$$
$$\mathcal{V}_{\mathrm{m}} = \left\{ v_{\mathrm{m}} \in \left[H^1\left(\Omega_{\mathrm{m}}\right)\right]^d,\ v_{\mathrm{m}}|_{\Gamma_{u\mathrm{m}}} = 0 \right\}. \qquad [1.37]$$

To account for the coupling in the penalty approach, the interface Dirichlet condition [1.36a] is weighted with its test counterpart, multiplied with a penalty parameter α_{pen} and added to the standard virtual work formulation of the uncoupled problem. In this way, we obtain the following variational formulation for the penalty coupling: find $(u_1, u_2) \in \mathcal{U}_1 \times \mathcal{U}_2$, such that:

$$\sum_{\mathrm{m}\in\{1,2\}} a_{\mathrm{m}}\left(u_{\mathrm{m}}, v_{\mathrm{m}}\right) + \alpha_{\mathrm{pen}} \int_{\Gamma} [\![u]\!] \cdot [\![v]\!] \mathrm{d}\Gamma$$
$$= \sum_{\mathrm{m}\in\{1,2\}} l_{\mathrm{m}}\left(v_{\mathrm{m}}\right), \quad \forall \left(v_1, v_2\right) \in \mathcal{V}_1 \times \mathcal{V}_2, \qquad [1.38]$$

where the standard bilinear form a_{m} and the linear form l_{m} associated with domain Ω_{m} ($\mathrm{m} \in \{1,2\}$) can be written as:

$$a_{\mathrm{m}}\left(u_{\mathrm{m}}, v_{\mathrm{m}}\right) = \int_{\Omega_{\mathrm{m}}} \varepsilon\left(v_{\mathrm{m}}\right) : C_{\mathrm{m}}\, \varepsilon\left(u_{\mathrm{m}}\right) \mathrm{d}\Omega \ ;$$
$$l_{\mathrm{m}}\left(v_{\mathrm{m}}\right) = \int_{\Omega_{\mathrm{m}}} v_{\mathrm{m}} \cdot f_{\mathrm{m}}^g \mathrm{d}\Omega + \int_{\Gamma_{F\mathrm{m}}} v_{\mathrm{m}} \cdot F_{\mathrm{m}}^g \mathrm{d}\Gamma \ ; \qquad [1.39]$$

and $[\![\cdot]\!]$ denotes the jump operator as follows:

$$[\![u]\!] = \left(u_1 - u_2\right); \quad [\![v]\!] = \left(v_1 - v_2\right). \qquad [1.40]$$

In the above equations, we note that we use the notations $\cdot$ and $:$ to refer to the scalar product of vector fields and of second-order tensor fields, respectively.

It is important to note that the accuracy of the method strongly depends on the choice of the penalty parameter. Indeed, a large value for α_{pen} is necessary to enforce the kinematic compatibility [1.36a]; however, making it too large makes the discrete coupled system ill-conditioned. A related drawback with this basic penalty method is that it is variationally inconsistent, in the sense that we cannot recover the interface conditions [1.36] of the strong form by using the weak form [1.38], especially the equilibrium of the tractions [1.36b]. Finally, it has to be noted that the stiffness operators of the sub-domains are merged together, which may complicate the separation of the computational resources between the different sub-domains. For all of these reasons, we will not consider this method in this book. Nevertheless, given its simplicity, we quote that the penalty approach remains of interest in IGA. Some efforts are currently performed to enhance such penalty techniques for isogeometric non-conforming coupling, in particular, with respect to the choice of the penalty parameter (see, for example, Herrema *et al.* (2019) and Coradello *et al.* (2020)).

1.5.3. *Mortar coupling*

A second option to weakly formulate the coupling problem [1.35]-[1.36] is to resort to the class of Lagrange multiplier methods. We note that the Lagrange multiplier approach is also often referred to as the Mortar approach in literature (Wohlmuth 2000; Brivadis *et al.* 2015; Dornisch *et al.* 2015; Zou *et al.* 2018; Wunderlich *et al.* 2019). We will use the two terminologies indifferently in the whole document. In this context, a mixed formulation is set up to impose the coupling constraints [1.36]. Classically, a single Lagrange multiplier $\lambda \in \mathcal{M}$ (where $\mathcal{M}$ is an appropriate space) is introduced, as the dual unknown, to represent both of the interface traction forces, i.e. $-\sigma_1 n_1 = \sigma_2 n_2 = -\lambda$ in equation [1.36b]. Then, the interface Dirichlet condition [1.36a] is imposed in a weak sense over Γ, using the Lagrange multiplier. This leads to the formulation of the following Lagrangian of the coupled problem:

$$\begin{aligned} L_{\text{basic}}\Big((u_1, u_2), \lambda\Big) = &\frac{1}{2} a_1(u_1, u_1) + \frac{1}{2} a_2(u_2, u_2) \\ &- l_1(u_1) - l_2(u_2) \\ &+ b(\lambda, u_1 - u_2), \end{aligned} \qquad [1.41]$$

where the bilinear form b is defined such that:

$$b(\mu, u) = \int_\Gamma \mu \cdot u \mathrm{d}\Gamma. \qquad [1.42]$$

With the above developments, we can finally obtain the classical Mortar coupling formulation of the reference problem as follows: find $(u_1, u_2, \lambda) \in \mathcal{U}_1 \times \mathcal{U}_2 \times \mathcal{M}$ such that:

$$a_1(u_1, v_1) + b(\lambda, v_1) = l_1(v_1), \quad \forall v_1 \in \mathcal{V}_1; \qquad [1.43a]$$

$$a_2(u_2, v_2) - b(\lambda, v_2) = l_2(v_2), \quad \forall v_2 \in \mathcal{V}_2; \qquad [1.43b]$$

$$b(\mu, u_1 - u_2) = 0, \quad \forall \mu \in \mathcal{M}. \qquad [1.43c]$$

Thanks to the additional Lagrange multiplier field, the formulation is time variationally consistent. Moreover, the practical advantage of such a mixed formalism is that it enables us to keep the stiffness operators associated with the sub-domains separated and unmodified. Indeed, the communication between the sub-domains is performed via the Lagrange multiplier only. This feature is the basis of the non-overlapping domain decomposition methods developed for high-performance computing on parallel computer architectures (Farhat and Roux 1991; Farhat *et al.* 2001; Stefanica 2001, 2005; Gosselet and Rey 2006; Kleiss *et al.* 2012). Drawing inspiration from these techniques, in Chapter 3 we will develop a family of efficient algorithms for non-conforming isogeometric multipatch analysis. As of now, it may be quoted that system [1.43] will not be solved directly, but in an iterative way through the resolution of several local symmetric positive definite systems. Using the same idea, the additional field also enables us to simply build non-invasive coupling algorithms for multiscale computations: several different numerical models along with their related numerical codes can be coupled iteratively with the exchange of interface data only (Passieux *et al.* 2013; Guguin *et al.* 2014; Duval *et al.* 2016; Guinard *et al.* 2018). This opportunity will be investigated in Chapter 2 for global/local IGA. However, formulation [1.43] represents a saddle point problem, meaning that special care may be required for the construction of the approximation space of $\mathcal{M}$ to avoid undesirable energy-free oscillations, due to the non-satisfaction of the *inf–sup* condition (see, for example, Ben Belgacem and Maday (1997) and Wohlmuth (2000, 2001) for the context of standard FEM and Brivadis *et al.* (2015) and Dittmann *et al.* (2019) for IGA). At this stage, we do not go into further detail and indicate that the Mortar coupling strategy will be discussed in Chapters 2 and 3.

NOTE.– In the context of Mortar approaches in IGA, it may be noted that a special interest is currently devoted to the dual Mortar resolution version (see, for example, Dornisch *et al.* (2015, 2017), Seitz *et al.* (2016), Zou *et al.* (2018) and Wunderlich *et al.* (2019)). The principle of this subclass of methods is to use dual-basis functions based on a biorthogonality concept to approximate $\mathcal{M}$, which allows for an easy elimination of the additional Lagrange multiplier DOF from the global system [1.43]. It results in a condensed coupled system that is positive definite and merges the stiffness operators of the two sub-domains. In this work, we follow another path for the resolution, as mentioned above: we benefit from the additional Lagrange multiplier field to develop high-performance computing and non-invasive tools, which appears original in the area of IGA.

1.5.4. *Nitsche coupling*

Unlike Mortar approaches, the coupling is established from a primal formulation in the Nitsche technique. This time, both of the interface Dirichlet and Neumann conditions [1.36] are enforced in a weak sense. More precisely, a connection between the Nitsche and Lagrange multiplier couplings can be made (see, for example, Fritz *et al.* (2004) and Bazilevs *et al.* (2012)). Starting with the Lagrange multiplier method, the idea to obtain the Nitsche method is to replace the Lagrange multiplier by the mean interface resultant force coming from the displacement. We therefore define the average of the stresses and of the virtual stresses as follows:

$$\begin{aligned}
&\{\sigma\} = \frac{1}{2}\left(\sigma_1 + \sigma_2\right) = \frac{1}{2}\left(C_1\, \varepsilon\left(u_1\right) + C_2\, \varepsilon\left(u_2\right)\right) \\
&\text{and } \{C\varepsilon\left(v\right)\} = \frac{1}{2}\left(C_1\, \varepsilon\left(v_1\right) + C_2\, \varepsilon\left(v_2\right)\right)
\end{aligned} \qquad [1.44]$$

and obtain the following Nitsche bilinear form:

$$\begin{aligned}
a_{\mathrm{N}}\Big((u_1, u_2), (v_1, v_2) \Big) = &\sum_{\mathrm{m}\in\{1,2\}} a_{\mathrm{m}}\left(u_{\mathrm{m}}, v_{\mathrm{m}}\right) \\
&- \int_\Gamma \{C\varepsilon\left(v\right)\}\, n_1 \cdot [\![u]\!] \mathrm{d}\Gamma \\
&- \int_\Gamma [\![v]\!] \cdot \{\sigma\}\, n_1 \mathrm{d}\Gamma,
\end{aligned} \qquad [1.45]$$

where we again use the jump operator [1.40]. This bilinear form finally needs to be enriched with a stabilization term to ensure the ellipticity of the boundary value problem. Denoting the stabilization parameter by α_{nit}, the stabilized variational formulation of the problem using the Nitsche approach can be written as follows: find $(u_1, u_2) \in \mathcal{U}_1 \times \mathcal{U}_2$, such that:

$$a_{\text{N}}\Big((u_1, u_2), (v_1, v_2) \Big) + \alpha_{\text{nit}} \int_{\Gamma} [\![u]\!] \cdot [\![v]\!] \mathrm{d}\Gamma$$

$$= \sum_{\text{m} \in \{1,2\}} l_{\text{m}}(v_{\text{m}}), \quad \forall (v_1, v_2) \in \mathcal{V}_1 \times \mathcal{V}_2. \qquad [1.46]$$

The main advantages of the Nitsche approach in comparison to other coupling techniques are its variational consistency, as well as the positive definiteness, symmetry and good conditioning properties directly obtained for the final coupled stiffness operator. In addition, the absence of auxiliary fields simplifies the theory. While in formulation [1.43], a suitable approximation space for the Lagrange multiplier needs to be chosen, the Nitsche approach [1.46] requires the choice of a suitable value for α_{nit}. It has been shown that good estimations of α_{nit} can be obtained, for example, by solving a generalized eigenvalue problem (Apostolatos *et al.* 2014b; Nguyen *et al.* 2014) or several local eigenvalue problems (Ruess *et al.* 2014; Jiang *et al.* 2015) over the interface. It is also worth mentioning at this stage that a non-symmetric version of the Nitsche method exists, which has the interest of being weakly stable without adding any stabilization terms (Schillinger *et al.* 2016a; Guo 2017; Hu *et al.* 2018). However, as with the penalty approach, the stiffness operators of the different sub-domains are merged together in the Nitsche coupling. This makes, at first glance, the derivation of efficient (parallel) algorithms and the use of different numerical codes for the resolution complicated, compared to Lagrange multiplier schemes. Furthermore, the true implementation of the method in contexts beyond standard solid elasticity may not be straightforward due to the emergence of the mean interface resultant stress [1.44]. For example, the formulation of the stress resultants transfer, for connecting two shells that intersect with an arbitrary angle, seems delicate (see Chapter 3). In the same idea, the definition of the virtual mean interface resultant stress is not obvious in the case of material nonlinearities (e.g. plasticity or damage). In Chapter 2, we will develop a new non-invasive coupling strategy based on the non-symmetric variant of the Nitsche methodology, which will result in a flexible and efficient tool to perform global/local isogeometric simulations.

NOTE.– It may be finally noted that with the Nitsche technique, the average operator in equation [1.44] can be defined more generally. Any convex combination can actually be chosen without violating the consistency of the formulation (Hansbo 2005; Bazilevs *et al.* 2012). As a result, we can use:

$$\begin{aligned}&\{\sigma\} = \gamma\sigma_1 + (1-\gamma)\sigma_2 = \gamma C_1\,\varepsilon\,(u_1) + (1-\gamma)C_2\,\varepsilon\,(u_2)\\ &\text{and } \{C\varepsilon\,(v)\} = \gamma C_1\,\varepsilon\,(v_1) + (1-\gamma)C_2\,\varepsilon\,(v_2)\\ &\text{with } \gamma \in [0,1]\,.\end{aligned} \qquad [1.47]$$

In particular, when there is an important contrast between the two materials to be coupled, it could be interesting to adjust the convex combination to counterbalance the material gap and therefore preclude the emergence of undesirable mesh locking phenomena (Sanders *et al.* 2012). Such treatment has been performed in the non-invasive algorithm proposed in Chapter 2, to properly introduce a stiff inclusion within a soft B-spline patch (see Bouclier and Passieux (2018)).

2

Non-invasive Coupling for Flexible Global/Local IGA

2.1. Brief introduction

This second chapter concerns the area of multiscale global/local isogeometric analysis of structures. The objective is to locally enrich a fixed, rigid (possibly multi-) patch isogeometric model by using a finer spline model (to perform local mesh refinement, or to introduce a geometrical detail such as a hole or a crack or even to model local nonlinearities such as plasticity and damage). Our approach to addressing this issue is based on the development of algorithms classified as "non-invasive". We show that this type of technology, which was originally introduced in the FEM framework to fulfill an industrial requirement, is of great interest for global/local IGA from a scientific point of view. The idea is to take the initial isogeometric model as the global model, in order to eliminate the need for costly spline re-parameterization procedures when integrating a non-conforming local model. From a more general point of view, this strategy also enables us to circumvent trimming, which constitutes one of the most important challenges to be overcome in IGA today (see section 1.4). Indeed, trimming inherently occurs when introducing an arbitrary local region within a global structured spline model. In Chapter 4 of this book, we will see that we can go even further by wisely using the developed strategy to achieve the optimal design of structural details simply and efficiently.

The present chapter is organized as follows: section 2.2 presents the origins of and reviews the standard non-invasive methodology; then section 2.3

further motivates the extension of the method to the field of IGA and outlines the scientific difficulties to be addressed; finally, section 2.4 describes the dedicated algorithm that we have built in response to these challenges.

2.2. The standard non-invasive strategy

The core idea of non-invasive global/local coupling has emerged in recent years to address a present-day industrial issue. The development of this class of methods was the object of a French research project supported by the French National Research Agency (ICARE ANR-12-MONU-0002, 01/2013–12/2016)[1]. This section establishes the non-invasive strategy as it comes in its initial FE form. We first highlight the motivations in an industrial context, and then derive the non-invasive coupling strategy with respect to the state-of-the-art in the field.

2.2.1. *Origin*

During the design, justification and certification process of industrial structures (such as an aircraft), design teams have to deal with different FE computer models, ranging from global to local representations. Global models assume simplifications of the geometry, kinematics and constitutive properties: their main purpose is to estimate loads in large sub-assemblies (e.g. fuselage sections, wings). Local models address greater complexity: their main goal is to capture the behavior of structural details up to rupture. To ensure the interaction between different scales, the most popular method in mechanical engineering is a top-down approach (often known as *submodeling* (Kelley 1982; Jara-Almonte and Knight 1988; Srinivasan *et al.* 1996; Cormier *et al.* 1999)), which is very common in commercial software. It consists of first carrying out a complete computation of the structure as a whole, and then using the obtained global solution to prescribe boundary conditions, as displacement (Kelley 1982) or load (Jara-Almonte and Knight 1988), on a more refined local model. This method presents the advantage of concentrating the computational effort on the zones that need it most. However, it is only applied in one direction (from global to local) without considering the repercussions of local, nonlinear effects on the global model,

1 Available at: http://www.institut-clement-ader.org/projets/icare/.

which may lead to severe disturbances of the overall balance of the structure (Cresta *et al.* 2007; Gendre *et al.* 2009).

As a result, it is necessary for most applications to use strong coupling to take into account the influence of the local model beyond its domain of definition. The conventional strategy is then to use monolithic coupling, i.e. the coupled solution is obtained through a direct computation (using a single direct solver). This requires a modification of the global model to remove certain elements and replace them with those of the local model. This reorganization of the global mesh (which takes many hours, days or even months to be constructed) is extremely invasive and has proved to be incompatible with industrial time cycles.

In response to this issue, a new class of global/local coupling methods, called *non-invasive* methods, has recently emerged in the computational structural mechanics community. Based on an idea put forward by Whitcomb (1991) and later formulated by Gendre *et al.* (2009) for modeling local plasticity, the strategy enables an FE model to be modified locally without having an impact on the corresponding numerical operators. More precisely, the method relies on an iterative process between global and local computations. The replacement of part of a global model by a more detailed local model is carried out precisely and non-invasively: the global model is never modified; only interface displacements and reaction forces are exchanged. This strategy has been successfully applied in many domains (see, for example, Gupta *et al.* (2012) and Passieux *et al.* (2013) for crack propagation, Chevreuil *et al.* (2013) and Nouy and Pled (2018) for taking into account localized uncertainties, Guguin *et al.* (2014, 2016) for non-invasive plate/3D coupling, Duval *et al.* (2016) and Gosselet *et al.* (2018) for domain decomposition solvers, Bettinotti *et al.* (2014) and Chantrait *et al.* (2014) for transient dynamics analysis, Oumaziz *et al.* (2018) for contact problems, Guinard *et al.* (2018) for real aeronautical structures and Blanchard *et al.* (2019) for viscoplastic structures).

2.2.2. *Non-invasive resolution of the coupling problem*

We now outline the construction of the standard non-invasive coupling strategy, i.e. the procedure that was initially developed in classical FEM throughout the French national project ICARE ANR. We also mention that the non-invasive algorithm presented hereafter is the one that has been considered preliminarily to connect IGA and FEM in multiscale simulations (see Bouclier *et al.* (2016) and Bouclier and Passieux (2022) (Chapter 2)), before being improved for the specific case of global/local IGA (see section 2.4).

2.2.2.1. *The global/local problem*

Let us first properly introduce the mechanical multiscale global/local problem that we aim to solve in this chapter. This problem results in a domain coupling problem, and thus it is similar to the one introduced in section 1.5.1. Analogous notations shown in Figure 1.22 are considered. The additional information that we take into account here is that we begin with the initial global (coarse) model of the structure without the local region (see Figure 2.1(a) (left) for illustration). More precisely, the model is characterized by a physical domain $\Omega_1 \subset \mathbb{R}^d$ (with $d = 2$ or 3 being the parametric dimension of the domain), which is divided into two disjoint open and bounded subsets Ω_{11} and Ω_{12}. These two non-overlapping sub-domains share a common interface, denoted by Γ, such that $\Omega_1 = \Omega_{11} \cup \Omega_{12} \cup \Gamma$ and $\Omega_{11} \cap \Omega_{12} = \varnothing$. A simple linear elastic law is adopted for the global model. We assume that such a constitutive behavior and the coarse spline discretization is sufficient for modeling and computation, except in the small region Ω_{12}, where a local phenomenon should be introduced. As a result, a local (more detailed) "sub-model", characterized by the domain Ω_2, is constructed to replace the global model in Ω_{12} (see Figure 2.1(a) (right)). The substitution of the local model within the global model is achieved by the interface Γ. The resulting global/local problem to be solved is a classical multi-domain problem in $\Omega_{11} \cup \Omega_2 \cup \Gamma$, with the global solution in Ω_{12} being discarded (see Figure 2.1(b)).

Following section 1.5.1, we denote by f_{11}^g and f_2^g the body forces applied over Ω_{11} and Ω_2, F_{11}^g and F_2^g the surfaces forces associated with boundaries $\Gamma_{F_{11}}$ and Γ_{F_2}, and u_{11}^g and u_2^g the displacements prescribed over boundaries $\Gamma_{u_{11}}$ and Γ_{u_2}, respectively (see Figure 2.1). Even though the standard non-invasive method applies for any (possibly nonlinear) local behavior, in the presentation in this section, we consider, for simplicity, that the local model is also linear elastic. Therefore, the strong form of the resulting coupling problem is given by equations [1.35] and [1.36], but with subscript 11 instead of subscript 1.

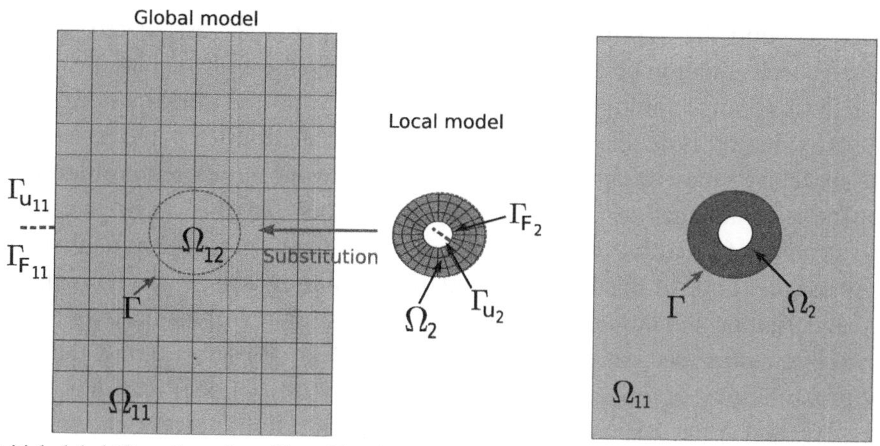

(a) Initial global ($\Omega_1 = \Omega_{11} \cup \Omega_{12} \cup \Gamma$) and local model domains. (b) Resulting coupling problem.

Figure 2.1. *Example of a global/local problem. The global model over sub-domain Ω_{12} is replaced by the finer local model of domain Ω_2 through the interface Γ, which enables a geometrical detail (a hole) to be integrated within the initial coarse model. For a color version of this figure, see www.iste.co.uk/bouclier/IGA.zip*

2.2.2.2. *Monolithic resolution*

The starting point in the derivation of the classical non-invasive strategy (i.e. in the sense of Passieux *et al.* (2013), Guguin *et al.* (2014) and Duval *et al.* (2016), for example) is to weakly formulate the coupling problem with a Lagrange multiplier approach. We thus resort to the mixed variational form [1.43] where, again, subscript 1 has to be replaced by subscript 11 to be consistent with our global/local modeling. Not taking care of non-invasiveness at the moment, we write the problem in the discrete setting below.

Regarding notations, we proceed as in Chapter 1 for discrete quantities: capital letters are used for matrices and lowercase letters are used for vectors (see, for example, equations [1.23] and [1.33]). We introduce the matrices $\mathbf{\Phi}_1$ and $\mathbf{\Phi}_2$ that collect the basis functions associated with the discretizations of the global and local models over Ω_1 and Ω_2, respectively (the same notation as given in equation [1.23], for example). At this stage, we remain very general regarding the discretization, i.e. the shape function matrices can be made of standard Lagrange polynomial, B-spline, NURBS or other basis functions. In addition, we denote by $\mathbf{\Phi}_{11}$ the matrix that gathers the restricted part to the sub-domain Ω_{11} of the shape functions of the global model. Following the principle of isoparametric elements, the basis functions into $\mathbf{\Phi}_{11}$ and $\mathbf{\Phi}_2$ are used to build the approximation subspaces $\mathcal{U}_{11}^h$ and $\mathcal{U}_2^h$ associated with $\mathcal{U}_{11}$ and $\mathcal{U}_2$ (see equation [1.37]), respectively. Finally, the approximation subspace for the Lagrange multiplier is denoted by $\mathcal{M}^h$ and the corresponding shape function matrix is denoted by $\mathbf{\Phi}_\lambda$. As outlined in section 1.5.3, the construction of space $\mathcal{M}^h$ may require special attention to avoid numerical problems (due to the non-satisfaction of the *inf–sup* condition). At this stage, we do not give more information on this point. Let us just say that a theoretical study is beyond the scope of this book; instead, our reasoning will be based on physical considerations and numerical experiments. This point will be discussed further in this chapter (see, in particular, section 2.4.1). By substituting the spline approximations of u_{11}, u_2 and λ in the weak form [1.43], and denoting by $\mathbf{u}_{11}$, $\mathbf{u}_2$ and $\boldsymbol{\lambda}$ the associated degrees of freedom (DOF) vectors, the following linear system is obtained:

$$\begin{bmatrix} \mathbf{K}_{11} & \mathbf{0} & \mathbf{C}_{11}^T \\ \mathbf{0} & \mathbf{K}_2 & -\mathbf{C}_2^T \\ \mathbf{C}_{11} & -\mathbf{C}_2 & \mathbf{0} \end{bmatrix} \begin{pmatrix} \mathbf{u}_{11} \\ \mathbf{u}_2 \\ \boldsymbol{\lambda} \end{pmatrix} = \begin{pmatrix} \mathbf{f}_{11} \\ \mathbf{f}_2 \\ \mathbf{0} \end{pmatrix}. \qquad [2.1]$$

The operators $\mathbf{K}_{11}$ (respectively, $\mathbf{f}_{11}$) and $\mathbf{K}_2$ (respectively, $\mathbf{f}_2$) are the classical stiffness matrices (respectively, vector forces) associated with subdomains Ω_{11} and Ω_2. $\mathbf{C}_{11}$ and $\mathbf{C}_2$ are the Mortar coupling operators that can be formally written as:

$$\mathbf{C}_{11} = \int_\Gamma \mathbf{\Phi}_\lambda \mathbf{\Phi}_{11}^T \mathrm{d}\Gamma \; ; \; \mathbf{C}_2 = \int_\Gamma \mathbf{\Phi}_\lambda \mathbf{\Phi}_2^T \mathrm{d}\Gamma. \qquad [2.2]$$

The resolution [2.1] of the global/local problem constitutes the classical monolithic approach: the resulting multiscale model of Figure 2.1(b) is computed directly using a single direct solver. This strategy can be characterized as invasive. Indeed, in addition to merging contributions from the global and local problems, it is important to note that the global operator $\mathbf{K}_{11}$ depends, at this stage, on the interface Γ or, in other words, on the shape of the local domain Ω_2. It is thus necessary to modify the initial global model to remove some of its elements, and possibly pieces of elements (see Figure 2.1). If the local detail grows up (during an optimization process, crack propagation, or expansion of damage or plasticity, for example), the situation is made even worse since not only the local operator $\mathbf{K}_2$, but also the global operator $\mathbf{K}_{11}$, has to be fully rebuilt and refactorized at each time step of the simulation.

2.2.2.3. *Non-invasive iterative resolution*

Rather than directly solving the system [2.1], the non-invasive strategy is based on an iterative exchange procedure. For the construction, we proceed in two steps: we first split the initial system in order to identify (in terms of boundary conditions applied on Γ) a Neumann and a Dirichlet problem over Ω_{11} and Ω_2, respectively; then, we apply the additivity of the integral with respect to the domain $\Omega_1 = \Omega_{11} \cup \Omega_{12} \cup \Gamma$ in order to recover the initial whole global model.

For the alternative resolution of [2.1], we can apply the following asymmetric algorithm: for the nth iteration, starting with initial guess $\boldsymbol{\lambda}^{(0)}$, we determine $\mathbf{u}_{11}^{(n)}$, $\mathbf{u}_2^{(n)}$ and $\boldsymbol{\lambda}^{(n)}$ such that:

1) Resolution of a Neumann problem over Ω_{11}:

$$\mathbf{K}_{11}\mathbf{u}_{11}^{(n)} = \mathbf{f}_{11} - \mathbf{C}_{11}^T\boldsymbol{\lambda}^{(n-1)}. \qquad [2.3]$$

2) Resolution of a Dirichlet problem over Ω_2:

$$\begin{bmatrix} \mathbf{K}_2 & -\mathbf{C}_2^T \\ -\mathbf{C}_2 & \mathbf{0} \end{bmatrix} \begin{pmatrix} \mathbf{u}_2^{(n)} \\ \boldsymbol{\lambda}^{(n)} \end{pmatrix} = \begin{pmatrix} \mathbf{f}_2 \\ -\mathbf{C}_{11}\mathbf{u}_{11}^{(n)} \end{pmatrix}. \qquad [2.4]$$

Looking carefully at the structure of the discrete equations, we can give a mechanical interpretation of the above fixed point strategy. Problem [2.4] actually constitutes a Dirichlet problem (in terms of boundary conditions applied on Γ): the second step of the iterative strategy [2.3]–[2.4] thus corresponds to the exchange of the interface displacement from the global to the local models ($\mathbf{C}_2\mathbf{u}_2^{(n)} = \mathbf{C}_{11}\mathbf{u}_{11}^{(n)}$ corresponds to $\mathbf{u}_2^{(n)}|_\Gamma = \mathbf{u}_{11}^{(n)}|_\Gamma$). Then, problem [2.3] consists of a Neumann problem (still in terms of boundary conditions applied on Γ): the interface traction force $\boldsymbol{\lambda}^{(n)}$ coming from the local model is applied to the global models. More information regarding the mechanical meaning of the standard non-invasive procedure can be found in Bouclier and Passieux (2022) (Chapter 2).

Even though two problems are now solved separately, the procedure [2.3]–[2.4] is still invasive because only the contributions over Ω_{11} of the global model are involved. To proceed further, we make use of the available continuous prolongation of the global solution from Ω_{11} to Ω_{12}. Denoting by $\mathbf{u}_1$ the DOF vector associated with domain Ω_1, we can write:

$$\mathbf{K}_1\mathbf{u}_1 = \overline{\mathbf{K}}_{11}\mathbf{u}_1 + \overline{\mathbf{K}}_{12}\mathbf{u}_1. \quad [2.5]$$

In the above equation, $\overline{\mathbf{K}}_{11}$ and $\overline{\mathbf{K}}_{12}$ are the extensions to Ω_1 of the classical stiffness matrices $\mathbf{K}_{11}$ and $\mathbf{K}_{12}$ related to sub-domains Ω_{11} and Ω_{12}, respectively. They literally contain the classical stiffness operators and are padded with zeros to make them the same dimension of $\mathbf{u}_1$. In addition, we define $\mathbf{f}_1 = \overline{\mathbf{f}}_{11} + \overline{\mathbf{f}}_{12}$ as the load vector associated with domain Ω_1. Equality [2.5] is used to expand the problem [2.3] from Ω_{11} to Ω_1. We finally end up with the following algorithm: for the nth iteration, starting with initial guesses $\boldsymbol{\lambda}^{(0)}$ and $\mathbf{u}_1^{(0)}$, we determine $\mathbf{u}_1^{(n)}$, $\mathbf{u}_2^{(n)}$ and $\boldsymbol{\lambda}^{(n)}$ such that:

1) Resolution of a Neumann problem over Ω_1:

$$\mathbf{K}_1\mathbf{u}_1^{(n)} = \mathbf{f}_1 - \mathbf{C}_1^T\boldsymbol{\lambda}^{(n-1)} + \mathbf{r}_{12}^{(n-1)}. \quad [2.6]$$

2) Resolution of a Dirichlet problem over Ω_2:

$$\begin{bmatrix} \mathbf{K}_2 & -\mathbf{C}_2^T \\ -\mathbf{C}_2 & \mathbf{0} \end{bmatrix} \begin{pmatrix} \mathbf{u}_2^{(n)} \\ \boldsymbol{\lambda}^{(n)} \end{pmatrix} = \begin{pmatrix} \mathbf{f}_2 \\ -\mathbf{C}_1\mathbf{u}_1^{(n)} \end{pmatrix} \quad [2.7]$$

We note that the Mortar operator $\mathbf{C}_1$ simply consists of the prolongation of the former operator $\mathbf{C}_{11}$ from Ω_{11} to Ω_1. $\mathbf{r}_{12}$ is introduced to denote the discrete reaction forces at Γ produced by the covered part Ω_{12} of the global model. This emerges to counterbalance the effect of this covered region, since it is not present in the reference coupling problem (see Figure 2.1(b)). It can be written at iteration $n-1$ as:

$$\mathbf{r}_{12}^{(n-1)} = \overline{\mathbf{K}}_{12}\mathbf{u}_1^{(n-1)} - \overline{\mathbf{f}}_{12}. \quad [2.8]$$

We can observe that in expression [2.8], only the DOF concerned with the interface Γ are not zero. It can also be emphasized that the fictitious prolongation of the global solution over Ω_{12} has no physical meaning. In Ω_2, only the local solution should be considered. An overview of the iterative procedure is shown in Figure 2.2. The procedure can be interpreted as a fixed point strategy aiming at ensuring the equilibrium of the interface reaction forces provided that the displacement is transferred at each iteration between the two models. Therefore, the convergence test usually considered to stop this algorithm can be written as:

$$\eta^{(n)} = \frac{||\mathbf{C}_1^T \boldsymbol{\lambda}^{(n)} + \mathbf{r}_{11}^{(n)}||_2}{\sqrt{||\mathbf{f}_1||^2 + ||\mathbf{f}_2||_2^2}}, \qquad [2.9]$$

where $\mathbf{r}_{11}^{(n)}$ characterizes the discrete reaction forces at Γ produced by the global model in the sub-domain Ω_{11}:

$$\mathbf{r}_{11}^{(n)} = \overline{\mathbf{K}}_{11} \mathbf{u}_1^{(n)} - \overline{\mathbf{f}}_{11}. \qquad [2.10]$$

The calculation [2.10] is performed from the already computed stiffness and vector force $\overline{\mathbf{K}}_{12}$ and $\overline{\mathbf{f}}_{12}$ (see equation [2.8]). In the rest of this chapter, we use the terminology "non-invasive residual" to denote the residual of an iterative non-invasive procedure.

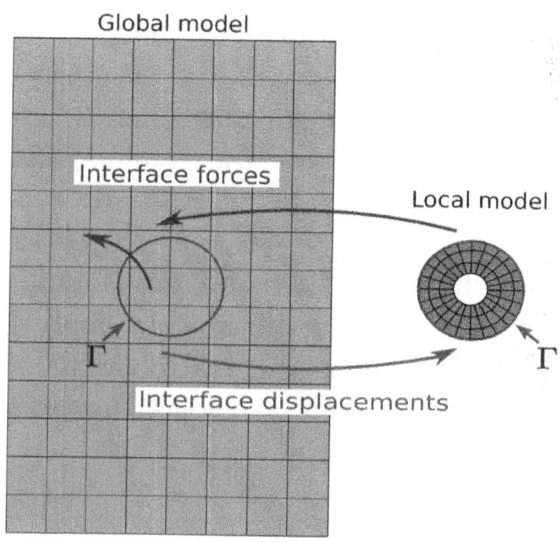

Figure 2.2. *The iterative non-invasive exchange procedure. For a color version of this figure, see www.iste.co.uk/bouclier/IGA.zip*

The algorithm shown in [2.6] and [2.7] constitutes what is commonly known as the non-invasive coupling strategy in the literature. The whole stiffness matrix of the initial global model is now considered without any modification. Since the global model is unmodified, the global stiffness operator can be assembled and factorized only once during the pre-processing step. Furthermore, with the global and local problems being solved alternatively and the interaction between the two models being restricted to the interface Γ, the formalism offers the possibility of combining different numerical codes with low implementation effort. The price to pay is the number of iterations of the fixed point solver. However, this can be drastically reduced by means of acceleration techniques, such as those based on Aitken's delta-squared method or a quasi-Newton method (see section 2.4.2 and, for example, Gendre *et al.* (2009), Liu *et al.* (2014) and Duval *et al.* (2016)), as will be demonstrated later in this chapter (see section 2.4.3).

NOTE.– The convergence of the standard iterative procedure [2.6] and [2.7] depends on the stiffness gap between the global (fictitious) model in Ω_{12} and the local model in Ω_2. Theoretically, this can be shown by rewriting the fixed point as a modified Newton algorithm where the approximation of the tangent matrix depends on the gap in the primal Schur complements between the models in Ω_{12} and Ω_2 (see, for example, Chevreuil *et al.* (2013), Duval *et al.* (2016) and Nouy and Pled (2018)). In particular, it can be proved that the standard strategy does not converge if the local model is stiffer than the global model. In this challenging situation, it is necessary to resort to the quasi-Newton acceleration technique to make the algorithm converge in a relatively low number of iterations (see, for example, Bouclier *et al.* (2016) and Duval *et al.* (2016).

NOTE.– As stated above, the method is by no means limited to a linear elastic local model. Indeed, as long as we are able to apply Dirichlet boundary conditions to the local problem and to compute (directly or in a post-processing step) the corresponding reaction forces, any local behavior can be considered.

2.3. Interest in the field of IGA

We believe that non-invasive coupling techniques are good candidates for the local enrichment of isogeometric models (i.e. to perform local mesh refinement, or to introduce a geometrical detail such as a hole or a crack or even to model local nonlinearities such as plasticity and damage). More than being attractive from an industrial engineering point of view, as in FEM, the non-invasive coupling idea appears to us completely relevant, from a scientific point of view, to improve the field of global/local structural IGA. More generally, it also enables us to nicely circumvent the trimming issue (see section 1.4.1), which inherently occurs when introducing an arbitrary local region within a global structured spline model. This section highlights the interest of building non-invasive (NURBS- or B-spline-based) isogeometric coupling methods. It also discusses the challenge of extending the above "standard" non-invasive method to the field of IGA.

2.3.1. *Global/local modeling in IGA*

As outlined in Chapter 1 (see section 1.2), (standard) multivariate spline bases comprise a rigid tensor product structure, which evidently constitutes a drawback to performing the global/local simulation. Indeed, we necessarily end up with a structured mesh in a B-spline or NURBS patch that precludes the simple modeling of local phenomena. More precisely, this unavoidably leads to the overlap of some global knot-span elements to allow a truly global mesh-independent local region to be incorporated for the modeling of any specific local behavior (e.g. introduction of a hole (Guo 2017), an inclusion (Nguyen *et al.* 2014), a crack (Bouclier *et al.* 2016)). From this point of view, Figure 2.1 seems to be appropriate to illustrate a situation that often occurs in IGA. In this context, we should consider the global model in Figure 2.1(a) as a B-spline patch and the local model as a NURBS patch. The analysis of trimmed isogeometric entities for the global model thus needs to be addressed, which is known in the IGA community not to be a trivial task (see section 1.4.1 and, for example, Marussig and Hughes (2018)). From the literature, three main strategies could be applied to numerically solve the resulting isogeometric global/local problem (see Figure 2.3).

2.3.1.1. *Spline re-parameterization*

The first strategy, already mentioned in section 1.4.1 (see Figure 1.16), consists of recovering the usual situation of boundary-fitted discretizations. In order to do so, a re-parameterization of the whole global/local isogeometric model is required, leading to the splitting of the new geometry into several patches with C^0 regularity at the boundaries (see, for example, Xu *et al.* (2013), Akhras *et al.* (2016) and Massarwi *et al.* (2019)). An illustration for the simple example of Figure 2.1 is given in Figure 2.3(a). For more complex models, this may entail considerable modeling and computational effort, which is often as complex and time-consuming as standard mesh generation and thus opposed to the core idea of IGA.

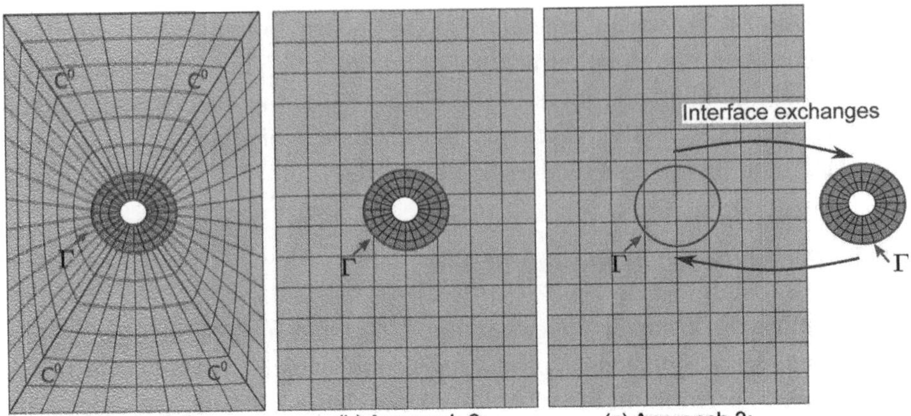

(a) Approach 1: Spline re-parametrization

(b) Approach 2: Unfitted domain

(c) Approach 3: Non-invasive strategy

Figure 2.3. *Three different strategies may be applied in IGA to numerically solve a multiscale global/local problem. For a color version of this figure, see www.iste.co.uk/bouclier/IGA.zip*

2.3.1.2. *Treatment of an unfitted spline domain*

Concurrently, a second approach is to resort to the family of immersed boundary (also referred to as the embedded domain, unfitted domain, finite cell or even CutFEM) methods that have been subjected to intensive studies over the last years (see Rank *et al.* (2012), Ruess *et al.* (2014), Burman *et al.* (2015) and Schillinger and Ruess (2015), to name a few). Unlike the first approach, the idea here is to simply use the resulting unfitted structured mesh for the interpolation of the global fields (see Figure 2.3(b)), while the global trimmed geometry is accurately captured by means of suitable quadrature rules for cut knot-span elements (Legrain 2013; Kudela *et al.* 2015; Nagy and Benson 2015; Fries and Omerovic 2016; Stavrev *et al.* 2016; Fries *et al.* 2017). This strategy is suitable and has proved to be highly efficient in the context of immersed finite elements. However, let us note that in the general case of a local region that may evolve during the simulation, for example, to carry out the shape optimization of local entities (see Chapter 4, especially section 4.7), or to model crack propagation, or the expansion of a plastic zone, etc., we expect several reassemblies and refactorizations of the resulting global/local stiffness operator during the multiresolution process. The left-hand side operator may also appear ill-conditioned depending on the encountered cut scenario (Burman *et al.* 2015; Dauge *et al.* 2015; Verhoosel *et al.* 2015; de Prenter *et al.* 2017).

2.3.1.3. *Non-invasive strategy*

The last strategy revolves around the concept of non-invasiveness, which was introduced in the previous section. The idea is to take the initial large-scale isogeometric (possibly multi-) patch domain to be enriched as the global model. This provides (i) simplicity, (ii) robustness and (iii) efficiency for the multiscale computation. Indeed, the global isogeometric model is never modified during the simulation (see Figure 2.3(c)), which eliminates (i) the need for costly spline re-parameterization procedures. In addition, the global stiffness operator remains (ii) well-conditioned regardless of the encountered cut configuration. Finally, the global stiffness operator is (iii) assembled and factorized only once in a pre-processing step. More specifically, if the local region evolves, thus leading to the resolution of a sequence of similar problems (in a multiresolution fashion (Gosselet *et al.* 2013; Néron *et al.* 2015)), the method meets its full potential, since the global stiffness operator remains the same; in particular, it ensures savings in computational time (Bouclier and Passieux 2018). Moreover, it should be mentioned that the flexibility of the strategy allows for the simple modeling of a variety of local behaviors. Indeed, since two different numerical codes can be considered to compute the global and local models, a linear isogeometric code can be used for the global simulation of the isogeometric patch, while a more elaborated code integrating specific complex behaviors can serve to incorporate an accurate local model. All of these advantages will be further accounted for (through numerical experiments) in section 2.4.3 and, later on, in Chapter 4.

2.3.2. *Challenges*

Before discussing the challenges involved in making the non-invasive coupling methodology suitable for IGA, let us return to the coupling terminology introduced in section 1.4.2. In the context of global/local IGA, the coupling should be performed at part or all of the boundary of the local model and within the global model. As a result, a *matching interface* is not expected, since this would have necessarily involved a spline re-parameterization of the global model (see approach 1 in Figure 2.3(a)), which is not desirable as explained above. We will then have either a *non-matching interface* when the boundary of the local model is aligned with the edges of the global knot-span elements, or a *non-conforming interface* when the boundary of the local model intersects with some global knot-span elements.

With a local model described using the standard FEM, considering only non-matching interfaces is sufficient for the global/local coupling since classical meshing procedures can be applied to make the boundary of the local mesh align with the edges of the global elements (see, for example, Guinard *et al.* (2018), which carries out multiscale global-shell/local-solid simulation for aeronautics and Bouclier and Passieux (2022) (Chapter 2), which performs hybrid global-IGA/local-FEM computations). This, however, is unthinkable in full IGA where both the global and local models are discretized by means of spline shape functions. Indeed, due to the tensor product structure of spline bases, building a tentative isogeometric transition mesh would again lead us to carrying out complex spline re-parameterization procedures. As a result, we need to truly address the situation of non-conforming interfaces, which requires us to (i) implement specific quadrature rules for the evaluation of integrals over pieces of coarse global elements (to obtain the interface reaction force $\mathbf{r}_{12}$, see equation [2.8]) and (ii) formulate a coupling procedure adapted to any cut scenarios encountered through the non-invasive multiresolution process.

We note that these challenges are also ubiquitous in the area of immersed boundary methods (see Figure 2.3(b) and, for example, Rank *et al.* (2012), Burman *et al.* (2015) and Schillinger and Ruess (2015)). We thus propose to draw inspiration from this active field of research to address these two technical issues. From this point of view, we may consider, for the numerical integration over pieces of elements, the recursive quadrature approach applied in the finite cell method (Ruess *et al.* 2014; Schillinger and Ruess 2015; Schillinger *et al.* 2016a), or more recent geometrically faithful quadratures (Legrain 2013; Kudela *et al.* 2015; Nagy and Benson 2015; Fries and Omerovic 2016; Stavrev *et al.* 2016; Fries *et al.* 2017), which can also be based on the moment fitting scheme (Müller *et al.* 2013; Joulaian *et al.* 2016; Hubrich *et al.* 2017; Hubrich and Düster 2019). It seems that the Nitsche coupling technology is preferred for properly connecting unfitted domains, since it circumvents the difficulty of choosing the compatible discrete dual space in the Mortar approach (see section 1.5). In what follows, we make use of such advanced fictitious domain techniques to build a new non-invasive algorithm that is able to handle all the non-conforming coupling scenarios encountered when performing global/local IGA.

NOTE.– When addressing domain coupling in IGA, we may also be interested in designing a coupling method that preserves higher regularity offered by the spline functions. This is of particular interest in the context of higher-order partial differential equations, such as for thin shell analysis in structural mechanics. In other words, the challenge would be to formulate a more regular coupling method that allows a smoother transition of the solution across the coupling interface. In Bouclier *et al.* (2017), a coupling scheme based on two sets of Lagrange multipliers that enables us to enforce a C^1 regularity at the interface was developed. Inspired from this work, we have built a new Mortar formulation adapted to the coupling of isogeometric Kirchhoff–Love shells (see Hirschler *et al.* (2019a, 2019c)). This formulation will be presented in Chapter 3. Let us finally note that subsequent studies from other authors with a similar objective have been performed very recently (see, for example, Dittmann *et al.* (2019), Schuß *et al.* (2019) and Paul *et al.* (2020)).

2.4. A robust algorithm for non-conforming global/local IGA

Let us now develop a numerical scheme that is more adapted for truly addressing the isogeometric non-invasive global/local coupling along a non-conforming interface. More precisely, the interaction between two non-conforming spline meshes is considered: one domain is the foreground and defines the local model; the other is the background and constitutes the global spline model to be enriched. In this part, we restrict ourselves to the case of linear elastic behavior for the local model (as well as for the global model). However, a note is added at the end of this section to discuss the extension for handling local nonlinearities. The shape function matrices $\mathbf{\Phi}_1$ and $\mathbf{\Phi}_2$ introduced in section 2.2.2 now read $\mathbf{R}_1$ and $\mathbf{R}_2$, respectively, to refer to NURBS (and thus also including B-spline) technologies. The work presented hereafter originates from Bouclier and Passieux (2018).

2.4.1. *Reference formulation: non-symmetric Nitsche coupling*

The new non-invasive algorithm that we propose is based on Nitsche-type coupling techniques and, in particular, the non-symmetric Nitsche variant. In the following, we give our motivations for our choice to resort to this coupling methodology and properly recall its formulation.

2.4.1.1. *Motivation*

Thanks to the introduction of the additional Lagrange multiplier field that represents the interface traction, it appears, at first glance, easier to make the two models communicate through interface exchanges while keeping the initial structure of the global and local stiffness operators. This may explain why the usual non-invasive strategy comes with a Mortar-based coupling method. However, as stated in section 1.5.3, the performance of the Mortar coupling is closely related to the compatibility between the approximation subspaces considered for the primal and dual variables. In the case of non-matching but conforming interfaces, there are some theoretical results that give indications for the construction of dual spaces, ensuring both stability and optimal accuracy of the coupling scheme (see, for example, Ben Belgacem and Maday (1997) and Wohlmuth (2000, 2001) for standard FEM, and Brivadis *et al.* (2015) and Dittmann *et al.* (2019) for IGA). For full global/local FEM or hybrid global-IGA/local-FEM, the non-invasive Mortar coupling approach seems to be well-adapted since the situation of non-matching coupling is sufficient for multiscale modeling (see the discussion in section 2.3.2). In these contexts, the common practice consists, for simplicity, of considering the trace along the coupling interface Γ of the basis functions of the local model for the construction of the dual space $\mathcal{M}^h$. With this practical choice, no disastrous instability problems have been reported in the field of non-invasive FEM coupling, even in challenging situations (Duval *et al.* 2016; Guinard *et al.* 2018).

In contrast, the case of non-conforming interfaces that arbitrarily cut the elements of the domains to be coupled makes the theoretical developments regarding Mortar approaches much more complicated and seems, in practice, to rule out many convenient choices for the discrete dual subspace. This is why Nitsche-based coupling is preferred in the field of immersed boundary methods. We recall that Nitsche formulations ensure variational consistency by introducing flux terms along the coupling interface, expressed by primal unknowns of the coupling domains (see section 1.5.4). Thus, the method is free of auxiliary fields, which simplifies the theory. Nevertheless, in its widespread symmetric version (see, for example, Annavarapu *et al.* (2012), Nguyen *et al.* (2014), Ruess *et al.* (2014), Buffa *et al.* (2019) and Elfverson *et al.* (2019)), additional stabilization terms are required to ensure coercivity of the formulation. Appropriate estimates of the stabilization parameters involved in solving an additional eigenvalue problem (whose resolution can be performed locally per element (Jiang *et al.* 2015)) are generally used, which may be delicate from an algorithmic viewpoint. Therefore, there has been an increasing interest in recent years in developing methods without such mesh-dependent stabilization parameters (Baiges *et al.* 2012; Kollmannsberger *et al.* 2015).

In this context, the non-symmetric version of the Nitsche method has recently reemerged in Dettmer *et al.* (2016), Schillinger *et al.* (2016a), Guo (2017) and Schöllhammer *et al.* (2019) for unfitted isogeometric methods. Originally introduced as a discontinuous Galerkin method (Oden *et al.* 1998; Rivière *et al.* 2001; Arnold *et al.* 2002), it is based on variationally consistent numerical flux conditions that are introduced in such a way that the criterion for stability is (weakly) satisfied. Therefore, it does not require the introduction of additional stabilization terms, in contrast to its symmetric counterpart, and thus its performance does not rely on an appropriate estimation of additional sensitive parameters. We also note that similar ideas are present in the literature under the name of skew-symmetric coupling (see, for example, Ladeveze *et al.* (2001), Chouly *et al.* (2017) and Hu *et al.* (2018)). The objective of this section is to further improve the initial non-invasive approach by extending it to incorporate the non-symmetric Nitsche coupling. The goal is to offer simplicity, accuracy and robustness, regardless of the non-conforming coupling scenarios encountered through the global/local multiresolution process. Indeed, we believe that the combination of the non-symmetric Nitsche coupling with the global/local iterative non-invasive strategy will result in an accurate and efficient tool to compute any evolution of a local model within a fixed spline model.

2.4.1.2. *Theory*

For a deeper insight into Nitsche methods, let us now introduce the non-symmetric coupling formulation from its initial discontinuous Galerkin framework. The development is based on Schillinger *et al.* (2016a), which provides a recent review of the non-symmetric Nitsche methodology applied to the Poisson problem. Here, we derive the scheme for the weak coupling of multiple domains in the context of linear elasticity. The obtained formulation will constitute the reference coupling for our new non-invasive strategy. The same notations as those used in section 2.2.2 for the coupling situation and in section 1.5 for the coupling equations are used in the following.

To begin with, we need to introduce, in addition to the functional spaces $\mathcal{U}_{\mathrm{m}}$ and $\mathcal{V}_{\mathrm{m}}$ for $\mathrm{m} \in \{11, 2\}$ (see equation [1.37]), the space for the stress solution and for the virtual stress field:

$$\Sigma_{\mathrm{m}} = \left\{\tau_{\mathrm{m}} \in H\left(\mathrm{div}, \Omega_{\mathrm{m}}\right), \ \tau_{\mathrm{m}} = \tau_{\mathrm{m}}^{T}\right\}, \tag{2.11}$$

where:

$$H\left(\mathrm{div}, \Omega_{\mathrm{m}}\right) = \left\{\tau_{\mathrm{m}} \in \left[L^2\left(\Omega_{\mathrm{m}}\right)\right]^{d \times d}, \ \mathrm{div}\left(\tau_{\mathrm{m}}\right) \in \left[L^2\left(\Omega_{\mathrm{m}}\right)\right]^{d}\right\}. \tag{2.12}$$

We note that equation [1.35c] is strongly enforced by considering the space $\mathcal{U}_{\mathrm{m}}$. To derive a Nitsche-type method, the first step consists of weakly imposing both the equilibrium and behavior of the structure. In order to do so, we multiply equations [1.35b] and [1.35a] by $C_{\mathrm{m}}^{-1}\tau_{\mathrm{m}}$ with $\tau_{\mathrm{m}} \in \Sigma_{\mathrm{m}}$ and by $v_{\mathrm{m}} \in \mathcal{V}_{\mathrm{m}}$, respectively, and perform integration by parts on each sub-domain Ω_{m}. Making use of the Neumann boundary condition [1.35d] as well, we find:

$$\int_{\Omega_{\mathrm{m}}} \tau_{\mathrm{m}} : C_{\mathrm{m}}^{-1}\sigma_{\mathrm{m}} \mathrm{d}\Omega = \int_{\partial\Omega_{\mathrm{m}}} \tau_{\mathrm{m}} n_{\mathrm{m}} \cdot u_{\mathrm{m}} \mathrm{d}\Gamma - \int_{\Omega_{\mathrm{m}}} \mathrm{div}\left(\tau_{\mathrm{m}}\right) \cdot u_{\mathrm{m}} \mathrm{d}\Omega \,; \tag{2.13}$$

$$\int_{\Omega_{\mathrm{m}}} \varepsilon\left(v_{\mathrm{m}}\right) : \sigma_{\mathrm{m}} \mathrm{d}\Omega = \int_{\Omega_{\mathrm{m}}} v_{\mathrm{m}} \cdot f_{\mathrm{m}}^{g} \mathrm{d}\Omega + \int_{\Gamma_{F\mathrm{m}}} v_{\mathrm{m}} \cdot F_{\mathrm{m}}^{g} \mathrm{d}\Gamma + \int_{\Gamma} v_{\mathrm{m}} \cdot \sigma_{\mathrm{m}} n_{\mathrm{m}} \mathrm{d}\Gamma. \tag{2.14}$$

We then discretize equations [2.13] and [2.14] by means of the approximation subspaces $\mathcal{U}_{\mathrm{m}}^{h} \subset \mathcal{U}_{\mathrm{m}}$, $\mathcal{V}_{\mathrm{m}}^{h} \subset \mathcal{V}_{\mathrm{m}}$ and $\Sigma_{\mathrm{m}}^{h} \subset \Sigma_{\mathrm{m}}$. We end up with the flux formulation (Arnold *et al.* 2002): find u_{m}^{h} and σ_{m}^{h} such that:

$$\int_{\Omega_{\mathrm{m}}} \tau_{\mathrm{m}}^{h} : C_{\mathrm{m}}^{-1}\sigma_{\mathrm{m}}^{h} \mathrm{d}\Omega = -\int_{\Omega_{\mathrm{m}}} \mathrm{div}\left(\tau_{\mathrm{m}}^{h}\right) \cdot u_{\mathrm{m}}^{h} \mathrm{d}\Omega + \int_{\Gamma_{u\mathrm{m}} \cup \Gamma_{F\mathrm{m}}} \tau_{\mathrm{m}}^{h} n_{\mathrm{m}} \cdot u_{\mathrm{m}}^{h} \mathrm{d}\Gamma + \int_{\Gamma} \tau_{\mathrm{m}}^{h} n_{\mathrm{m}} \cdot \hat{u}_{\mathrm{m}}^{h} \mathrm{d}\Gamma \,; \tag{2.15}$$

$$\int_{\Omega_{\mathrm{m}}} \varepsilon\left(v_{\mathrm{m}}^{h}\right) : \sigma_{\mathrm{m}}^{h} \mathrm{d}\Omega = \int_{\Omega_{\mathrm{m}}} v_{\mathrm{m}}^{h} \cdot f_{\mathrm{m}}^{g} \mathrm{d}\Omega + \int_{\Gamma_{F\mathrm{m}}} v_{\mathrm{m}}^{h} \cdot F_{\mathrm{m}}^{g} \mathrm{d}\Gamma + \int_{\Gamma} v_{\mathrm{m}}^{h} \cdot \hat{\sigma}_{\mathrm{m}}^{h} n_{\mathrm{m}} \mathrm{d}\Gamma \,; \tag{2.16}$$

where the numerical fluxes $\hat{u}_{\mathrm{m}}^h$ and $\hat{\sigma}_{\mathrm{m}}^h$ are the approximations of the displacement and stress, respectively, on the interface Γ. To obtain the non-symmetric Nitsche formulation, the following expressions for the numerical fluxes are considered:

$$\hat{u}_{\mathrm{m}}^h = \frac{3}{2} u_+^h - \frac{1}{2} u_-^h \ ; \qquad [2.17]$$

$$\hat{\sigma}_{\mathrm{m}}^h = \frac{1}{2}\left(\sigma_+^h + \sigma_-^h\right); \qquad [2.18]$$

where the subscript $+$ is used to denote a variable of sub-domain Ω_{m}, whereas the subscript $-$ refers to a variable of the neighboring sub-domain. This new notation is convenient here since it allows us to characterize a sub-domain and its neighbor independently of $\mathrm{m} \in \{11, 2\}$. To make use of expressions [2.17] and [2.18], we now apply the following integration by parts formula:

$$-\int_{\Omega_{\mathrm{m}}} \operatorname{div}\left(\tau_{\mathrm{m}}^h\right) \cdot u_{\mathrm{m}}^h \mathrm{d}\Omega = \int_{\Omega_{\mathrm{m}}} \tau_{\mathrm{m}}^h : \varepsilon\left(u_{\mathrm{m}}^h\right) \mathrm{d}\Omega - \int_{\partial\Omega_{\mathrm{m}}} \tau_{\mathrm{m}}^h n_{\mathrm{m}} \cdot u_{\mathrm{m}}^h d\Gamma. \qquad [2.19]$$

Considering equation [2.15], replacing its second term by the right-hand side terms of equation [2.19], choosing $\tau_{\mathrm{m}}^h = C_{\mathrm{m}}\varepsilon\left(v_{\mathrm{m}}^h\right)$ on its left-hand side term, and inserting the flux approximation [2.17], we then obtain:

$$\int_{\Omega_{\mathrm{m}}} \varepsilon\left(v_{\mathrm{m}}^h\right) : \sigma_{\mathrm{m}}^h \mathrm{d}\Omega = \int_{\Omega_{\mathrm{m}}} \varepsilon\left(v_{\mathrm{m}}^h\right) : C_{\mathrm{m}}\, \varepsilon\left(u_{\mathrm{m}}^h\right) \mathrm{d}\Omega + \int_{\Gamma} C_{\mathrm{m}}\, \varepsilon\left(v_{\mathrm{m}}^h\right) n_{\mathrm{m}} \cdot \frac{1}{2}\left(u_+^h - u_-^h\right) \mathrm{d}\Gamma. \qquad [2.20]$$

Finally, by making use of the flux approximation [2.18] into [2.16], relating the result to the left-hand side of [2.20] and summing over the two sub-domains Ω_{11} and Ω_2, we end up with the non-symmetric Nitsche coupling formulation: find $\left(u_{11}^h, u_2^h\right) \in \mathcal{U}_{11}^h \times \mathcal{U}_2^h$ such that:

$$\sum_{\mathrm{m}\in\{11,2\}} a_{\mathrm{m}}\left(u_{\mathrm{m}}^h, v_{\mathrm{m}}^h\right) + \int_{\Gamma} \left\{C\varepsilon\left(v^h\right)\right\} n_{11} \cdot [\![u^h]\!] \mathrm{d}\Gamma - \int_{\Gamma} [\![v^h]\!] \cdot \left\{\sigma^h\right\} n_{11} \mathrm{d}\Gamma = \sum_{\mathrm{m}\in\{11,2\}} l_{\mathrm{m}}\left(v_{\mathrm{m}}^h\right),\ \forall \left(v_{11}^h, v_2^h\right) \in \mathcal{V}_{11}^h \times \mathcal{V}_2^h, \qquad [2.21]$$

where the bilinear form a_{m} and the linear form l_{m} can be written as in [1.39], and the jump and average operators are defined as in [1.40] and [1.44], respectively.

In view of carrying out a consistency study of the coupling method, another formulation can be written. Indeed, replacing u^h_{m} by $u_{\mathrm{m}} \in \mathcal{U}_{\mathrm{m}}$ and v^h_{m} by $v_{\mathrm{m}} \in \mathcal{V}_{\mathrm{m}}$ in the primal formulation [2.21], integrating by parts, and bringing all the terms on the left-hand side, we can write:

$$\sum_{\mathrm{m}\in\{11,2\}} -\int_{\Omega_{\mathrm{m}}} v_{\mathrm{m}} \cdot (\mathrm{div}(\sigma_{\mathrm{m}}) + f^g_{\mathrm{m}})\, \mathrm{d}\Omega + \int_{\Gamma_{F_{\mathrm{m}}}} v_{\mathrm{m}} \cdot (\sigma_{\mathrm{m}}\, n_{\mathrm{m}} - F^g_{\mathrm{m}})\, \mathrm{d}\Gamma \tag{2.22a}$$

$$+ \int_\Gamma \frac{1}{2} C_{11} \varepsilon\,(v_{11})\, n_{11} \cdot (u_{11} - u_2)\, \mathrm{d}\Gamma + \int_\Gamma v_{11} \cdot \frac{1}{2}\,(\sigma_{11} - \sigma_2)\, n_{11} \mathrm{d}\Gamma \tag{2.22b}$$

$$+ \int_\Gamma \frac{1}{2} C_2 \varepsilon\,(v_2)\, n_{11} \cdot (u_{11} - u_2)\, \mathrm{d}\Gamma + \int_\Gamma v_2 \cdot \frac{1}{2}\,(\sigma_{11} - \sigma_2)\, n_{11} \mathrm{d}\Gamma = 0. \tag{2.22c}$$

Expression [2.22] will be of interest in highlighting the physical meaning of the developed algorithm (see below). We note, at this stage, that terms [2.22a] represent the mechanical equilibrium of each sub-domain Ω_{m}, while terms [2.22b] and [2.22c] correspond to the non-symmetric Nitsche enforcement of the coupling conditions [1.36]. It can be proved that the formulation is variationally consistent with respect to the strong form of the initial problem [1.35] and [1.36], the criterion for stability is (weakly) satisfied and the method converges with optimal rates in the strain energy norm (Rivière 2008; Burman 2011).

2.4.2. *A Nitsche-based non-invasive algorithm*

We now address the non-invasive resolution of the non-symmetric Nitsche coupling. After clarifying the classical monolithic approach in this situation, the proposed non-invasive algorithm is constructed. Finally, a reformulation of the iterative procedure is performed to enlighten the mechanical and numerical properties of the new algorithm.

2.4.2.1. *Monolithic resolution*

For ease of understanding of the next developments, let us first note that equation [2.21] can be written as follows: find $u_{11}^h \in \mathcal{U}_{11}^h$ and $u_2^h \in \mathcal{U}_2^h$ such that:

$$a_{11}\left(u_{11}^h, v_{11}^h\right) - \int_\Gamma v_{11}^h \cdot \left\{\sigma^h\right\} n_{11} \mathrm{d}\Gamma + \int_\Gamma \frac{1}{2} C_{11} \varepsilon\left(v_{11}^h\right) n_{11} \cdot [\![u^h]\!] \mathrm{d}\Gamma$$
$$= l_{11}\left(v_{11}^h\right), \quad \forall v_{11}^h \in \mathcal{V}_{11}^h ; \qquad [2.23]$$

$$a_2\left(u_2^h, v_2^h\right) - \int_\Gamma v_2^h \cdot \left\{\sigma^h\right\} n_2 \mathrm{d}\Gamma - \int_\Gamma \frac{1}{2} C_2 \varepsilon\left(v_2^h\right) n_2 \cdot [\![u^h]\!] \mathrm{d}\Gamma$$
$$= l_2\left(v_2^h\right), \quad \forall v_2^h \in \mathcal{V}_2^h. \qquad [2.24]$$

The resolution [2.23] and [2.24] of the coupling problem constitutes the classical monolithic approach: in other words, this is the counterpart of [2.1] using the non-symmetric Nitsche theory.

2.4.2.2. *Making the Nitsche method non-invasive*

To begin with, we propose, for the resolution of [2.23]–[2.24], the following algorithm that enables us to split the initial multi-domain problem into one problem over Ω_{11} and a second problem over Ω_2. Hereafter, we voluntarily omit specifying the functional spaces in which the solutions are searched for, to avoid making the equations more cumbersome. For the nth iteration, starting with initial guesses $u_{11}^{h(0)}$ and $u_2^{h(0)}$, we determine $u_{11}^{h(n)}$ and $u_2^{h(n)}$ such that:

1) Resolution of a problem over Ω_{11}:

$$a_{11}\left(u_{11}^{h(n)}, v_{11}^h\right) = l_{11}\left(v_{11}^h\right) + \int_\Gamma v_{11}^h \cdot \left\{\sigma^{h(n-1)}\right\} n_{11} \mathrm{d}\Gamma$$
$$- \int_\Gamma \frac{1}{2} C_{11} \varepsilon\left(v_{11}^h\right) n_{11} \cdot [\![u^{h(n-1)}]\!] \mathrm{d}\Gamma. \qquad [2.25]$$

2) Resolution of a problem over Ω_2:

$$a_2\left(u_2^{h(n)}, v_2^h\right) - \int_\Gamma v_2^h \cdot \frac{1}{2} \sigma_2^{h(n)} n_2 \mathrm{d}\Gamma + \int_\Gamma \frac{1}{2} C_2 \varepsilon\left(v_2^h\right) n_2 \cdot u_2^{h(n)} \mathrm{d}\Gamma$$
$$= l_2\left(v_2^h\right) + \int_\Gamma v_2^h \cdot \frac{1}{2} \sigma_{11}^{h(n)} n_2 \mathrm{d}\Gamma + \int_\Gamma \frac{1}{2} C_{2\varepsilon\left(v_2^h\right)} n_2 \cdot u_{11}^{h(n)} \mathrm{d}\Gamma. \qquad [2.26]$$

A mechanical interpretation of the resulting iterative strategy will be given later in this section. Nevertheless, we may already note that formulation [2.26] corresponds to the non-symmetric Nitsche weak imposition of the coupling conditions only for the sub-domain Ω_2. Indeed, integrating by parts, and bringing all the terms on the left-hand side, we find:

$$0 = -\int_{\Omega_2} v_2 \cdot \left(\mathrm{div}(\sigma_2^{(n)}) + f_2^g\right) \mathrm{d}\Omega + \int_{\Gamma_{F_2}} v_2 \cdot \left(\sigma_2^{(n)}\, n_2 - F_2^g\right) \mathrm{d}\Gamma \tag{2.27}$$

$$+\int_{\Gamma} \frac{1}{2} C_2 \varepsilon\left(v_2\right) n_2 \cdot \left(u_2^{(n)} - u_{11}^{(n)}\right) \mathrm{d}\Gamma + \int_{\Gamma} v_2 \cdot \frac{1}{2}\left(\sigma_2^{(n)} - \sigma_{11}^{(n)}\right) n_2 \mathrm{d}\Gamma. \tag{2.28}$$

The above equation clearly shows that, at each iteration of the proposed algorithm, the interface term [2.22c] is recovered from the initial coupling formulation.

We now switch to the matrix form for ease of reading with respect to the implementation. From procedure [2.25] and [2.26], the only thing that remains to be done to truly reach a non-invasive algorithm consists of recovering the initial global model over Ω_1 in equation [2.25]. To this end, we may simply repeat the procedure of equation [2.5]. This leads us to build the following non-invasive algorithm based on the non-symmetric Nitsche coupling. Reusing the notations introduced in section 2.2, the algorithm reads: for the nth iteration, starting with initial guesses $\mathbf{u}_1^{(0)}$ and $\mathbf{u}_2^{(0)}$, we determine $\mathbf{u}_1^{(n)}$ and $\mathbf{u}_2^{(n)}$ such that:

1) Resolution over Ω_1:

$$\begin{aligned}\mathbf{K}_1 \mathbf{u}_1^{(n)} = \mathbf{f}_1 &- \left(\overline{\mathbf{K}}_{11}^{N^T} - \overline{\mathbf{K}}_{11}^{N}\right) \mathbf{u}_1^{(n-1)} \\ &- \left(\overline{\mathbf{K}}_{21}^{N^T} - \overline{\mathbf{K}}_{12}^{N}\right) \mathbf{u}_2^{(n-1)} + \mathbf{r}_{12}^{(n-1)}.\end{aligned} \tag{2.29}$$

2) Resolution over Ω_2:

$$\left[\mathbf{K}_2 + \left(\overline{\mathbf{K}}_{22}^{N^T} - \overline{\mathbf{K}}_{22}^{N}\right)\right] \mathbf{u}_2^{(n)} = \mathbf{f}_2 - \left(\overline{\mathbf{K}}_{12}^{N^T} - \overline{\mathbf{K}}_{21}^{N}\right) \mathbf{u}_1^{(n)}. \tag{2.30}$$

The different blocks of the Nitsche coupling operator are computed as follows:

$$\overline{\mathbf{K}}_{11}^{N} = \int_{\Gamma} -\frac{1}{2}(\mathbf{Z}_{11}\,\mathbf{H}_1\,\mathbf{B}_1)^T\mathbf{R}_1^T \mathrm{d}\Gamma\,; \quad [2.31a]$$

$$\overline{\mathbf{K}}_{12}^{N} = \int_{\Gamma} \frac{1}{2}(\mathbf{Z}_{11}\,\mathbf{H}_1\,\mathbf{B}_1)^T\mathbf{R}_2^T \mathrm{d}\Gamma\,; \quad [2.31b]$$

$$\overline{\mathbf{K}}_{21}^{N} = \int_{\Gamma} -\frac{1}{2}(\mathbf{Z}_{11}\,\mathbf{H}_2\,\mathbf{B}_2)^T\mathbf{R}_1^T \mathrm{d}\Gamma\,; \quad [2.31c]$$

$$\overline{\mathbf{K}}_{22}^{N} = \int_{\Gamma} \frac{1}{2}(\mathbf{Z}_{11}\,\mathbf{H}_2\,\mathbf{B}_2)^T\mathbf{R}_2^T \mathrm{d}\Gamma\,; \quad [2.31d]$$

where $\mathbf{B}_1$ and $\mathbf{B}_2$ represent the standard spline strain–displacement operators, $\mathbf{H}_1$ and $\mathbf{H}_2$ model the Hooke constitutive law, and the operator $\mathbf{Z}_{11}$ is introduced to perform the product between the stress tensor and the outward unit normal to Ω_{11} (see Nguyen *et al.* (2014) and Ruess *et al.* (2014) for more details regarding the construction of such operators).

As discussed in section 2.3.2, specific quadrature rules need to be implemented in the case of a non-conforming global/local interface to evaluate $\mathbf{r}_{12}$ (see equation [2.8]). In order to do so, a large amount of techniques, again, taken from immersed boundary methods, can be used. In this work, for simplicity and robustness, we perform the recursive quadrature approach used in FCM (Ruess *et al.* 2014; Schillinger and Ruess 2015; Schillinger *et al.* 2016a). The principle of this integration technique is illustrated in 2D in Figure 2.4. In the case of a trimmed element (i.e. in this work involving a global knot-span element cut by the coupling interface), it is replaced by four equally spaced sub-cells. The process is then repeated for all sub-cells that are cut by the interface, and so on until a predefined partitioning level (5 in Figure 2.4) is reached. Finally, every created sub-cell that lies in the true domain is equipped with $(p+1)\times(q+1)$ Gauss points (with p and q being the polynomial degree of the spline functions in the two parametric directions). For the sub-cells of the last level that remain cut by the interface, only the Gauss points within the true domain are taken into account in the quadrature rule. The quadtree approach can be easily adjusted to 3D using octrees. The reader must be careful not to confuse sub-cells with spline elements: sub-cells are only built for the quadrature rule and not for the discretization of the mechanical fields which obviously rely on the (initial) isogeometric mesh (see Figure 2.4(top)).

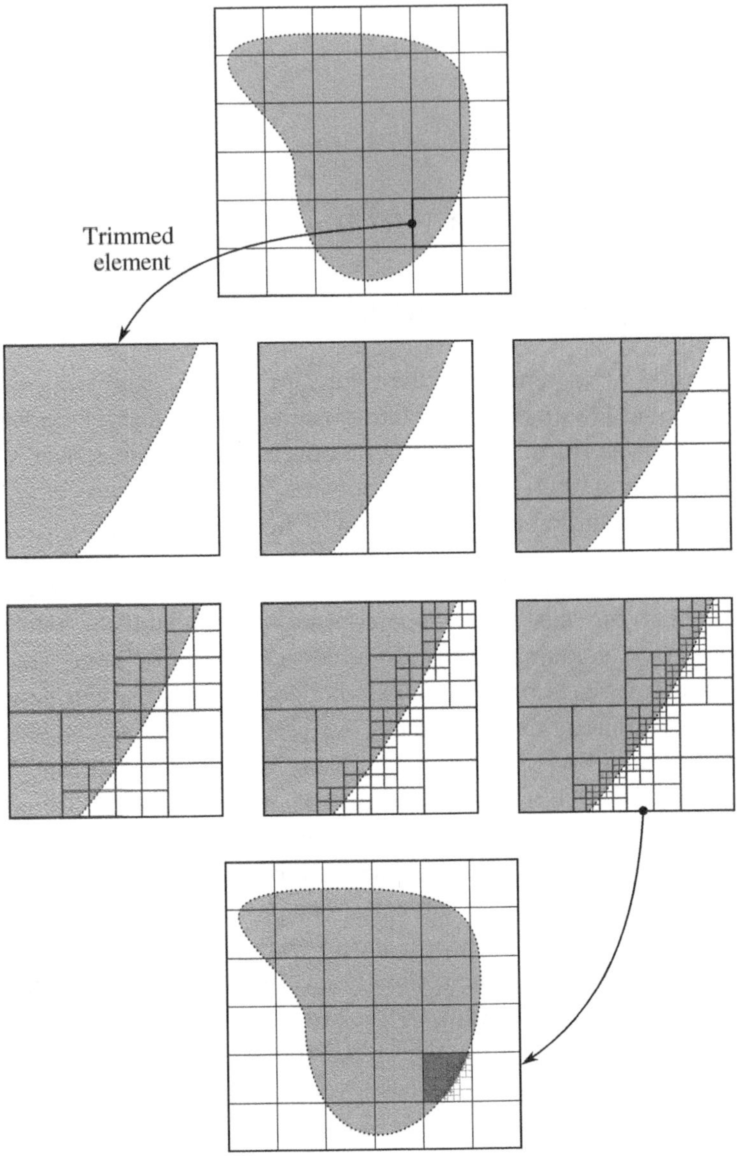

Figure 2.4. *Principle of the recursive quadrature approach used in FCM (Ruess* et al. *2014; Schillinger and Ruess 2015; Schillinger* et al. *2016a)*

NOTE.– Let us again emphasize at this stage that the issue of ill-conditioning that is usually encountered in immersed methods (Dauge *et al.* 2015; Verhoosel *et al.* 2015; de Prenter *et al.* 2017) is circumvented with the algorithm [2.29] and [2.30]. Indeed, the initial global stiffness operator, which is symmetric and well-conditioned, is given in [2.29] (instead of the cut global model over Ω_{11}), which provides robustness to the method. According to us, this feature is very attractive and may open the way for solving the ill-conditioning issue in a more general unfitted finite element context (Burman *et al.* 2015; Schillinger and Ruess 2015).

2.4.2.3. *Incremental formulation and mechanical interpretation*

Another incremental formulation of the non-invasive algorithm [2.29]–[2.30] can be derived for a better understanding of the numerical procedure. This alternative formulation is given in the following. The emphasis is given to the mechanical interpretation of the formulation, but not the rigorous theoretical analysis of the algorithm.

To begin with, let us denote by $\mathbf{r}^N$ the Nitsche-based discrete reaction forces at Γ applied to the global model in equation [2.29]. We have, at iteration $n-1$:

$$\mathbf{r}^{N(n-1)} = \left(\overline{\mathbf{K}}_{11}^{N\,T} - \overline{\mathbf{K}}_{11}^{N}\right)\mathbf{u}_1^{(n-1)} + \left(\overline{\mathbf{K}}_{21}^{N\,T} - \overline{\mathbf{K}}_{12}^{N}\right)\mathbf{u}_2^{(n-1)}. \quad [2.32]$$

Then, let us note the following equality that corresponds to the discrete equilibrium of the global model at iteration $n-1$:

$$\mathbf{K}_1\mathbf{u}_1^{(n-1)} = \mathbf{f}_1 + \mathbf{r}_{11}^{(n-1)} + \mathbf{r}_{12}^{(n-1)}, \quad [2.33]$$

where $\mathbf{r}_{11}$ characterizes the discrete reaction forces at Γ produced by the global model in the sub-domain Ω_{11} (see equation [2.10]), i.e.:

$$\mathbf{r}_{11}^{(n-1)} = \overline{\mathbf{K}}_{11}\mathbf{u}_1^{(n-1)} - \overline{\mathbf{f}}_{11}. \quad [2.34]$$

Now, performing the subtraction between equations [2.29] and [2.33], we arrive at:

$$\mathbf{K}_1\left(\mathbf{u}_1^{(n)} - \mathbf{u}_1^{(n-1)}\right) = -\left(\mathbf{r}^{N(n-1)} + \mathbf{r}_{11}^{(n-1)}\right). \quad [2.35]$$

As a result, the procedure [2.29] and [2.30] can be rewritten as follows: for the nth iteration, starting with the initial guess $\mathbf{u}_1^{(0)}$, we determine $\mathbf{u}_1^{(n)}$ such that:

$$\mathbf{u}_1^{(n)} = \mathbf{u}_1^{(n-1)} - \mathbf{K}_1^{-1}\mathbf{g}\left(\mathbf{u}_1^{(n-1)}\right), \qquad [2.36]$$

where the application $\mathbf{g}$ is defined as:

$$\mathbf{g}\left(\mathbf{u}_1^{(n-1)}\right) = \mathbf{r}^{N(n-1)} + \mathbf{r}_{11}^{(n-1)}, \qquad [2.37]$$

provided the local solution $\mathbf{u}_2^{(n-1)}$ in $\mathbf{r}^{N(n-1)}$ is computed from $\mathbf{u}_1^{(n-1)}$ through the system [2.30] (expressed at iteration $n-1$).

Formulation [2.36] shows that the iterative procedure simply consists of a fixed point aiming at solving $\mathbf{g} = \mathbf{0}$. The non-invasive residual [2.9] thus becomes here:

$$\eta^{(n)} = \frac{||\mathbf{r}^{N(n)} + \mathbf{r}_{11}^{(n)}||_2}{\sqrt{||\mathbf{f}_1||_2^2 + ||\mathbf{f}_2||_2^2}}. \qquad [2.38]$$

Moreover, writing the continuous version of equation [2.37], we literally find:

$$g\left(u_1^{(n-1)}\right) = \int_\Gamma \frac{1}{2} C_{11}\varepsilon\left(v_{11}\right) n_{11} \cdot \left(u_1^{(n-1)} - u_2^{(n-1)}\right) \mathrm{d}\Gamma$$
$$- \int_\Gamma v_1 \cdot \frac{1}{2}\left(\sigma_1^{(n-1)} + \sigma_2^{(n-1)}\right) n_{11}\mathrm{d}\Gamma + \int_\Gamma v_1 \cdot \sigma_1^{(n-1)} n_{11}\mathrm{d}\Gamma, \qquad [2.39]$$

which corresponds to the consistency term [2.22b] of the initial Nitsche coupling formulation. Therefore, the proposed iterative strategy aims to ensure the coupling conditions for the sub-domain Ω_{11} in a weak non-symmetric Nitsche sense. In addition, recalling that the other interface consistency term [2.22c] is enforced at each iteration of the non-invasive procedure through the local problem [2.30] (see the term [2.28] that corresponds to the coupling conditions related to the domain Ω_2), we ensure that if the proposed algorithm is convergent, all the interface terms of the initial Nitsche coupling formulation [2.22] are recovered. In other words, if the developed non-invasive algorithm is convergent, it leads to the same solution as the one computed from the monolithic approach [2.23] and [2.24]. This numerical procedure is original and its mechanical interpretation is quite different from the standard non-invasive strategy that aims to recover the interface force equilibrium, provided that the Dirichlet conditions (over Γ) are transmitted, at each iteration, from the global to the local problem by means of a Lagrange multiplier field (see section 2.2.2).

Although different from the standard non-invasive strategy in its mechanical interpretation, the algorithm can still be interpreted as a modified Newton method prescribed on $\mathbf{g} = \mathbf{0}$ (see equation [2.36]). As a result, the application of the same acceleration techniques as the ones usually performed in the context of non-invasive Lagrange multiplier coupling is straightforward. In this work, we consider two acceleration methods: the Aitken dynamic relaxation and the quasi-Newton update. The first technique, based on Aitken's delta-squared formulation (Irons and Tuck 1969; Küttler and Wall 2008), proved to be very simple, cheap and of relatively good acceleration performance, especially when the global and local models are rather close (e.g. local mesh refinement, local plasticity). With a slightly more computational overhead, the second method enables us to further speed up the algorithm and make it convergent even in a challenging situation. The inverse of the symmetric rank-one update (Khalfan *et al.* 1993; Kelley and Sachs 1998) (which converges towards $\nabla \mathbf{g}^{-1}$) is computed in a non-invasive fashion by means of the Sherman–Morrison formula. For more information on these acceleration techniques, the interested reader is advised to consult (Gendre *et al.* 2009; Duval *et al.* 2016).

NOTE.– It should be noted at this stage that more recent strategies, such as those based on the reformulation of the algorithm by means of a conjugate gradient procedure (Gosselet *et al.* 2018), can also be considered to further improve the numerical efficiency of the algorithm. In this spirit, we may note that other splits could have been proposed (such as replacing $\mathbf{u}_1^{(n)}$ by $\mathbf{u}_1^{(n-1)}$ and/or $\overline{\mathbf{K}}^N \mathbf{u}_2^{(n)}$ by $\overline{\mathbf{K}}^N \mathbf{u}_2^{(n-1)}$ in equation [2.30]). In combination with the use of Krylov solvers, this would enable us to perform in a similar way to domain decomposition methods (Farhat and Roux 1991; Le Tallec *et al.* 1991; Gosselet and Rey 2006).

2.4.3. *Validation*

To assess the performance of the developed algorithm [2.29]–[2.30], we now present a series of numerical experiments in two-dimensional linear elasticity. The plane stress assumption is considered for all test cases. In this section, the units are not specified: all the input data and the results are given in a consistent unit system. The examples include different discretization scenarios that cover non-matching and non-conforming couplings along straight and curved interfaces. More precisely, the strategy is validated through three different benchmarks, involving local mesh refinement, introduction of a geometrical detail and coupling at a bi-material interface. From here onwards, the mesh composed of N elements along the first length and M elements along the second length will be denoted as N-by-M. The k-refinement strategy (see section 1.2.3 and Cottrell *et al.* (2007)) is performed so that the maximum level of regularity (i.e. $C^{(p-1)}$ for a polynomial degree p) is ensured across the interior knots. In the illustrations, we keep the notations introduced in section 2.2.2; in particular, the domain $\Omega_1 = \Omega_{11} \cup \Omega_{12} \cup \Gamma$ characterizes the global isogeometric model and the local model of the domain Ω_2 is expected to replace the global model in the sub-domain Ω_{12}.

2.4.3.1. *Curved beam subjected to end shear*

The first example consists of a curved beam subjected to end shear. The global geometry is perfectly generated using a single NURBS patch composed of only one quadratic element. The problem, together with its global/local discretization, is shown in Figure 2.5(a). A constant horizontal displacement of $u_0 = 0.01$ units is prescribed over the lower beam boundary. An analytical solution is available for the problem in Zienkiewicz *et al.* (2005). In the upper half of the structure, the global NURBS model is meant to be replaced by a more refined (along the radial direction) NURBS model. Quadratic NURBS functions are considered for the discretization of both models. The situation of non-matching meshes to be coupled is investigated here: the global/local interface is aligned with the edges of the global knot-span elements. As a result, no specific quadrature rules need to be implemented for the computation of $\mathbf{r}_{12}$ (see equation [2.8]). To integrate over the interface, we divide the straight line into 1D sub-elements by combining the trace on the interface of the two meshes.

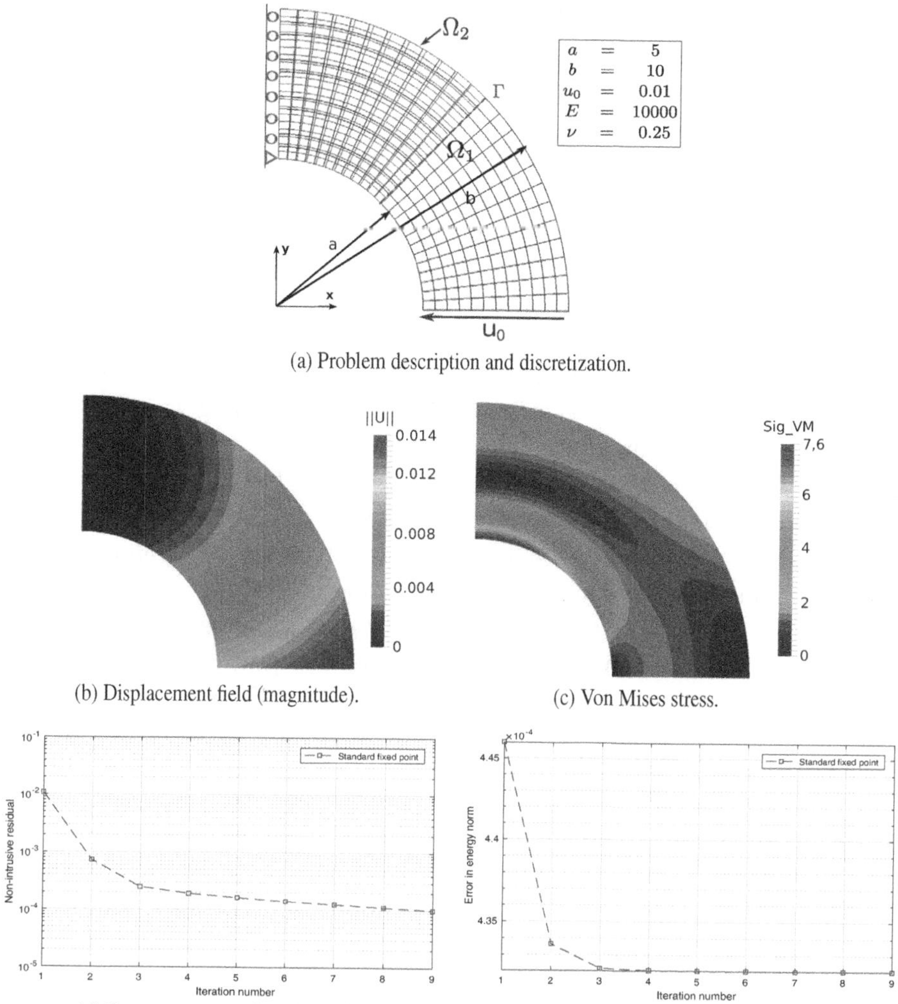

(a) Problem description and discretization.

(b) Displacement field (magnitude).

(c) Von Mises stress.

(d) Convergence: non-invasive residual.

(e) Convergence: error in energy norm.

Figure 2.5. *Global/local non-invasive Nitsche analysis of the curved beam problem (NURBS mesh of quadratic 24- (circumferential direction) by-12 (radial direction) elements for Ω_1 (12-by-12 for Ω_{11}) and NURBS mesh of quadratic 12-by-20 elements for Ω_2). For a color version of this figure, see www.iste.co.uk/bouclier/IGA.zip*

The results obtained by performing the algorithm [2.29] and [2.30] with the discretization of Figure 2.5(a) are grouped in Figure 2.5(b)–(e). More precisely, the iterative algorithm is converged to obtain the solution, which is plotted in Figure 2.5(b) and (c) in terms of displacement and von Mises stress, respectively. We note that it is the coupled solution in $\Omega_{11} \cup \Omega_2$ that is mapped (the fictitious prolongation of the global solution over Ω_{12} is not represented). The solution is in good agreement with Zienkiewicz *et al.* (2005). In particular, there is no visible error concentration around the coupling interface Γ. Figure 2.5(d) and (e) enable the convergence of the iterative strategy to be appreciated: first, in terms of the non-invasive residual (equation [2.38]) and then in terms of the error on the displacements in the energy norm (which is taken with respect to the analytical solution). No acceleration techniques are implemented here since the standard fixed point is already efficient. The non-invasive residual tends to zero and the error on the displacements reaches an asymptotic value (corresponding to the NURBS finite element approximation). This accounts for the convergence of the algorithm. More importantly, it may actually be noted from Figure 2.5(e) that it takes three iterations to reach the stagnation of the NURBS finite element error. This means that it takes only three iterations for the coupling error to be negligible compared to that of the discretization. It seems that a criterion based on a coupling error of 10^{-4} is slightly too demanding. To go further, a suitable error estimator should be implemented to stop the iterative algorithm at the right time (Duval *et al.* 2018; Tirvaudey *et al.* 2020). Such an estimator is beyond the scope of this work, and we choose to stop the algorithm at 10^{-4} in the following.

The convergence of the method with the refinement of the mesh is studied. For this purpose, Figure 2.6 shows the convergence curve for the error in the energy norm. To refine the coupled solution, we consider the meshes indicated in Table 2.1. For comparison purposes, the convergence curve of the equivalent single-patch solution is also plotted in Figure 2.6. The convergence curves are plotted with respect to the equivalent mesh size normalized by the mesh size of the equivalent coarsest mesh. Finally, the solution of the standard non-invasive strategy [2.6] and [2.7], based on a simple Lagrange multiplier coupling, is added to the graphic. For the construction of the dual space, we follow the common practice in non-invasive coupling, i.e. we take the trace along the interface of the local shape functions (see the discussion at the beginning of section 2.4.1). The results show that optimal rates of convergence, along with optimal accuracy, are achieved on this test case, both with the new Nitsche-based and standard Lagrange-based non-invasive strategies. This means that the accuracy of the derivative quantities is preserved with the two methodologies. In the context of solid mechanics where, from an engineering point of view, the accuracy of the stresses is crucial for the structural design and optimization, this property is of primary importance. However, we emphasize that here we are considering a simple coupling situation of non-matching meshes. As a result, the choice adopted for the approximation of the dual variable in the standard approach is suitable in this particular situation, which gives the full accuracy of the resulting mixed method. As will be illustrated in the next examples, this property of the Lagrange multiplier method will deteriorate when addressing truly non-conforming interfaces.

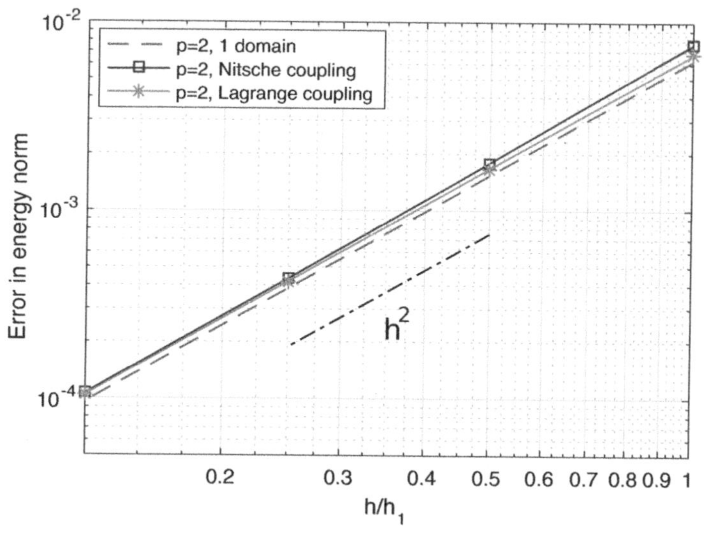

Figure 2.6. *Convergence of the error in the energy norm for the curved beam problem. For a color version of this figure, see www.iste.co.uk/bouclier/IGA.zip*

Total number of elements	Single-patch mesh	Non-matching meshes ($\Omega_{11} \cup \Omega_2$)
24	6×4	$3 \times 3 \cup 3 \times 5$
96	12×8	$6 \times 6 \cup 6 \times 10$
384	24×16	$12 \times 12 \cup 12 \times 20$
1536	48×32	$24 \times 24 \cup 24 \times 40$

Table 2.1. *Meshes considered to study the convergence behavior with respect to mesh refinement for the curved beam problem*

2.4.3.2. *Infinite plate with a circular hole*

In the next step, we study the performance of the proposed methodology for the introduction of a geometrical detail in a given spline model. As an example, we consider the popular test case of an infinite plate with a circular hole under in-plane tension. The geometry, material, boundary conditions and analytical stress solution (Saad 2009) are shown in Figure 2.7. The hole constitutes the geometrical detail to be incorporated. The discretization of the problem following the developed non-invasive strategy is shown in Figure 2.8(a). A regular B-spline mesh is used for the plate without the hole (domain Ω_1) and a circular refined ring-shaped NURBS domain is constructed, as the local model, for an accurate representation of the stress concentration around the hole. Therefore, this time, the case of non-conforming coupling along a curved interface is investigated. Overall, six levels of recursive quadrature sub-cells (Schillinger and Ruess 2015) are used to ensure accuracy with regard to the evaluation of $\mathbf{r}_{12}$ (see Figure 2.8(b) and (c)). This precaution holds for all the remaining examples in this chapter. Finally, to integrate over the non-conforming interface, we refine the NURBS curve irrespective of the underlying global and local meshes.

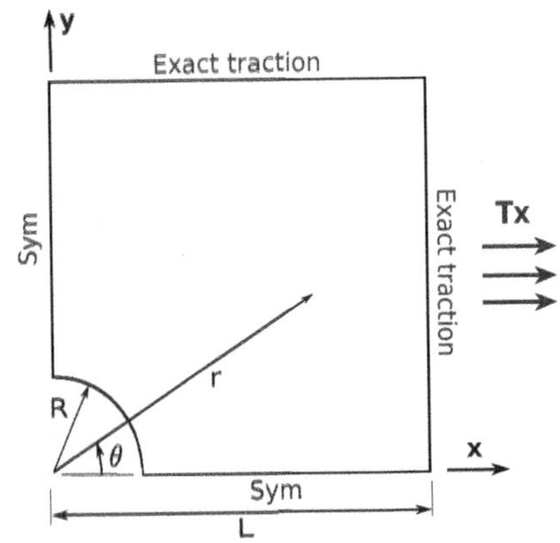

Parameters:

$$
\begin{array}{lcl}
L & = & 4 \\
R & = & 1 \\
E & = & 10^5 \\
\nu & = & 0.3 \\
T_x & = & 10
\end{array}
$$

Exact solution:

$$
\begin{aligned}
\sigma_{rr}(r,\theta) &= \frac{T_x}{2}\left(1-\frac{R^2}{r^2}\right)+\frac{T_x}{2}\left(1+3\frac{R^4}{r^4}-4\frac{R^2}{r^2}\right)\cos(2\theta) \\
\sigma_{\theta\theta}(r,\theta) &= \frac{T_x}{2}\left(1+\frac{R^2}{r^2}\right)-\left(1+3\frac{R^4}{r^4}\right)\cos(2\theta) \\
\sigma_{r\theta}(r,\theta) &= -\frac{T_x}{2}\left(1+2\frac{R^2}{r^2}-3\frac{R^4}{r^4}\right)\sin(2\theta)
\end{aligned}
$$

Figure 2.7. *Infinite plate with a hole: description and data of the problem. For a color version of this figure, see www.iste.co.uk/bouclier/IGA.zip*

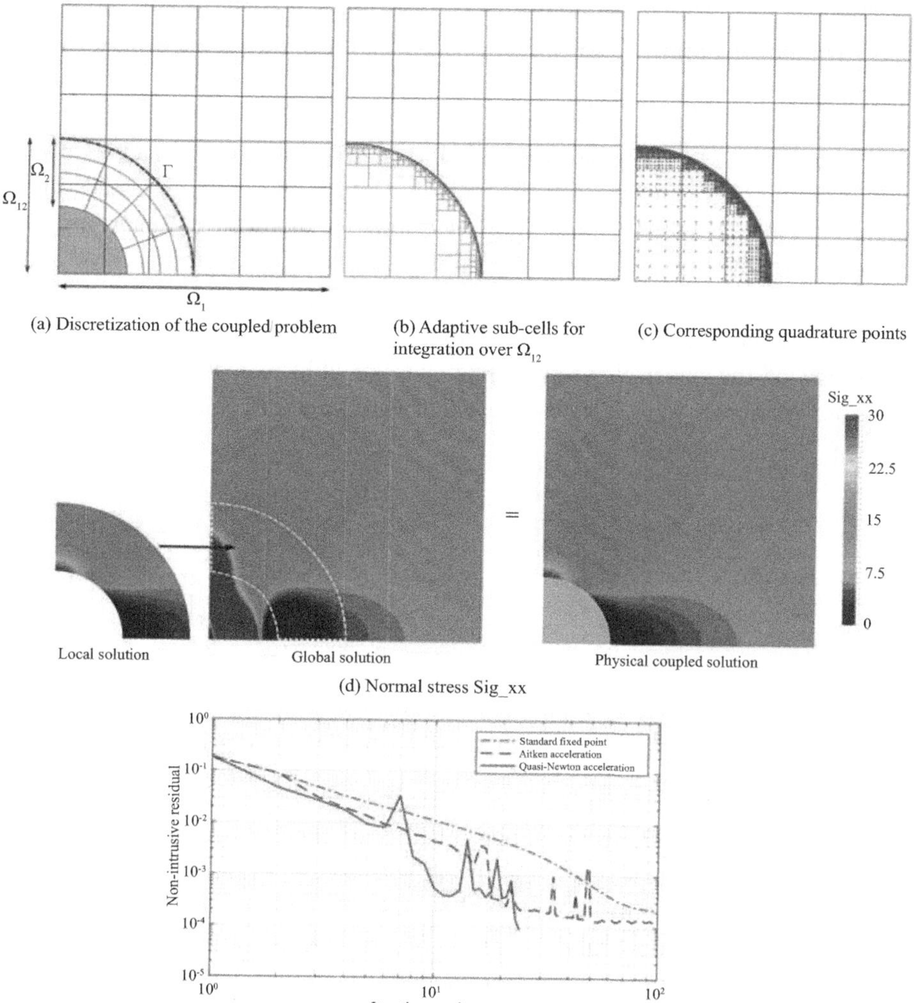

Figure 2.8. *Global/local non-invasive Nitsche analysis of the infinite plate with a hole (rectangular B-spline mesh of cubic 6-by-6 elements for Ω_1, recursive Gaussian quadrature (Schillinger and Ruess 2015) with six levels of sub-cells for Ω_{12} and circular NURBS mesh of cubic 4-by-4 elements for Ω_2). For a color version of this figure, see www.iste.co.uk/bouclier/IGA.zip*

The results obtained with the algorithm [2.29]–[2.30] and the discretization of Figure 2.8(a) are shown in Figure 2.8(d) and (e) in terms of stress and non-invasive convergence, respectively. Removing the smooth non-physical fictitious prolongation in Ω_{12}, and replacing part of it by the local solution in Ω_2, the resulting global/local solution appears to be smooth and in good agreement with the results obtained using NURBS-fitted discretizations (Hughes *et al.* 2005; Cottrell *et al.* 2009). Regarding the convergence of the proposed algorithm, the quasi-Newton acceleration technique seems to be necessary for this test case. This may be expected here since the stiffness gap between the domains Ω_{12} and Ω_2 is significant. Taking advantage of the quasi-Newton update, we are able to make the number of iterations relatively low: a residual of 10^{-4} is obtained in about 20 iterations.

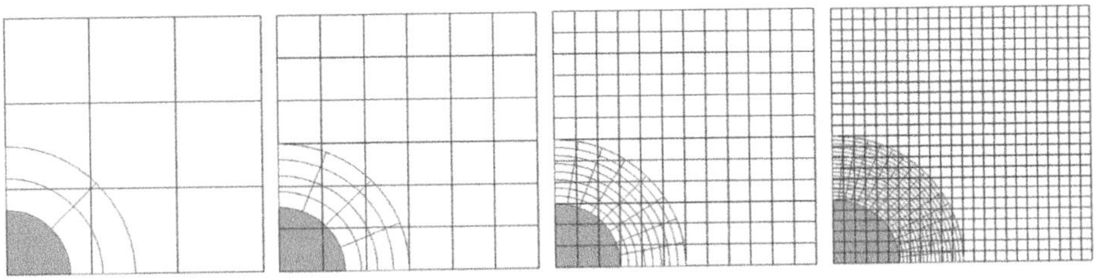

Figure 2.9. *Sequence of meshes considered for the convergence study. For a color version of this figure, see www.iste.co.uk/bouclier/IGA.zip*

The finite element convergence of the method when applied to this non-conforming scenario is then investigated (see Figures 2.9 and 2.10). Second-, third- and fourth-degree spline shape functions are considered for both the global and local models. The different meshes of Figure 2.9 are built for the numerical study. They correspond to several uniform refinements starting from a mesh of 3-by-3 B-spline elements for the global model and 2-by-2 NURBS elements for the local model. The corresponding convergence curves obtained with the error in the energy norm are plotted in Figure 2.10(a) and (b) for the proposed Nitsche-based and standard Lagrange multiplier-based strategies, respectively. Unlike what was experienced for the previous curved beam problem, the Lagrange multiplier approach failed to reproduce the optimal convergence rates, especially when increasing the polynomial degree of the spline shape functions. The reason for this poor accuracy may be due to the non-conforming situation encountered, which rules out the practical choice adopted for the construction of the dual subspace. Conversely, without any additional specific treatment, the proposed non-symmetric Nitsche version seems to achieve the optimal convergence rates, which demonstrates that the methodology does not interfere with the accuracy of the spline functions, regardless of the coupling scenario encountered. Note that for the finest quartic coupling discretization, the error level is so low that it may be deteriorated by rounding errors.

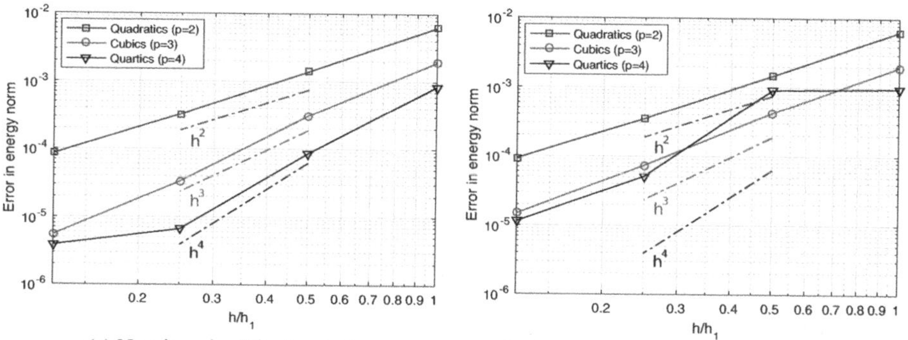

(a) Non-invasive Nitsche strategy. (b) Standard non-invasive Lagrange strategy.

Figure 2.10. *Convergence curves in energy norm for the infinite plate with a hole. For a color version of this figure, see www.iste.co.uk/bouclier/IGA.zip*

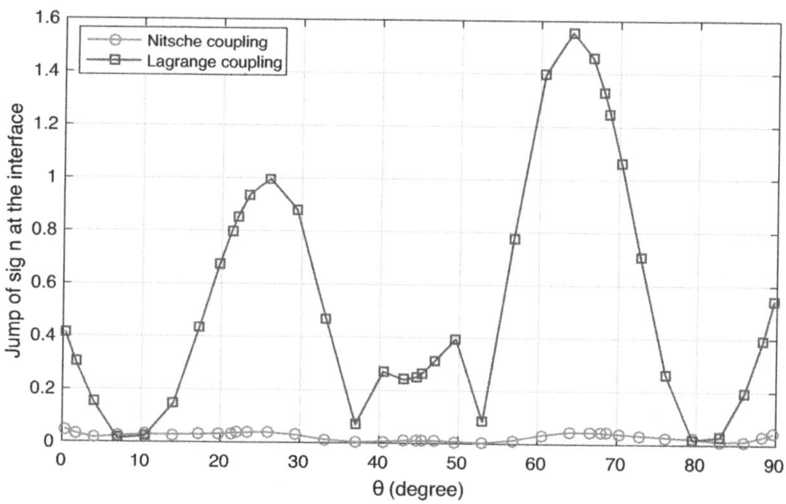

Figure 2.11. *Jump of the interface tractions along Γ (computed as $||\sigma_1 n_{11} + \sigma_2 n_2||_\Gamma$) for the infinite plate with a hole (rectangular B-spline mesh of quartic 6-by-6 elements for Ω_1 and circular NURBS mesh of quartic 4-by-4 elements for Ω_2). For a color version of this figure, see www.iste.co.uk/bouclier/IGA.zip*

For completeness, the jump of the traction forces across Γ with respect to the parametric coordinate along the interface is plotted in Figure 2.11 in the case of a Nitsche-based as well as a Lagrange multiplier-based non-invasive strategy. The second quartic discretization is considered. We observe that the Nitsche-based strategy enables us to greatly reduce the numerical jump in order to reproduce equation [1.36b], whereas significant oscillations appear in the case of a Lagrange multiplier-based procedure, which confirms the superiority of the developed strategy when addressing non-invasive non-conforming coupling.

2.4.3.3. *Plate with a central inclusion*

With the last example, the situation of a non-conforming, curved and bi-material interface is investigated, considering the modeling of a plate with a central inclusion subjected to constant in-plane tension (see Figure 2.12(a)). The Young modulus is chosen to be 100 times larger for the inclusion than for the plate ($E_i = 100E_p$). The Poisson ratios are the same. Regarding the discretization, a regular quadratic B-spline grid composed of 28-by-40 elements is used for the plate, while a circular quadratic NURBS mesh with 48- (circumferential direction) by-12 (radial direction) elements is constructed to model the inclusion. For the integration over intersected entities, we keep the recursive quadrature approach and refine the NURBS interface irrespective of the underlying models. The interest of such a test case, involving an overlapping mesh, both finer and stiffer than the underlying grid, is that it may display mesh locking, as shown in the context of classical FEM (see, for example, Béchet *et al.* (2009) and Lew and Buscaglia (2008)). For comparison purposes, a refined conforming FE solution is computed using the classical four-node element implemented in the industrial software ABAQUS (see Figure 2.12(b)).

The results in terms of von Mises stress are presented in Figure 2.13. The quasi-Newton acceleration technique was used to make the algorithm converge and about 50 iterations were necessary to handle such a challenging situation. Due to the contrast in Young's moduli at the coupling interface, the same treatment as proposed in Sanders *et al.* (2012) for the symmetric Nitsche coupling in the context of standard immersed FEM is applied to the

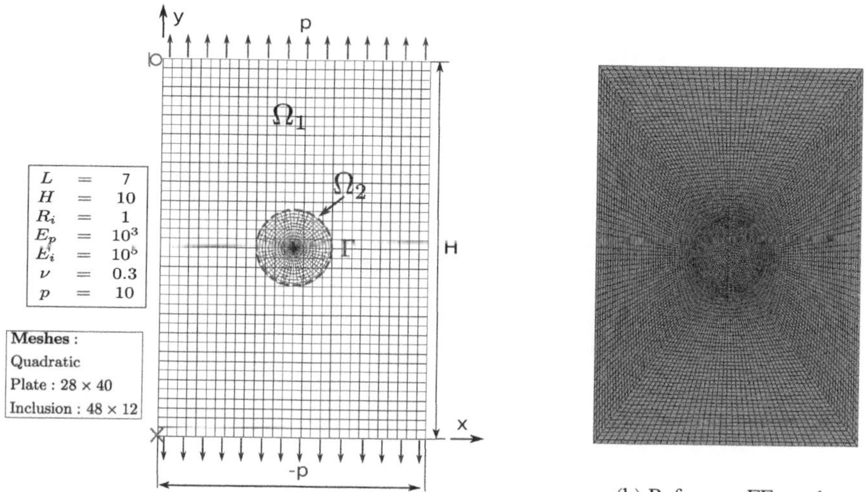

(a) Spline global/local non-invasive problem.

(b) Reference FE mesh (from ABAQUS).

Figure 2.12. *Plate with a central inclusion: description and discretization of the global/local problem. (Note: here, $E_1 = E_p$ stands for the Young modulus associated with the plate, while $E_2 = E_i$ denotes the Young modulus of the inclusion). For a color version of this figure, see www.iste.co.uk/bouclier/IGA.zip*

proposed non-invasive non-symmetric Nitsche coupling. It consists of resorting to the general form [1.47] of the average operator and to take:

$$\gamma = \frac{E_2}{E_1 + E_2}. \qquad [2.40]$$

to counterbalance the material gap at the interface. Thanks to this precaution, we are able to accurately recover the physical stress state with the developed non-invasive strategy (the stiffer behavior of the inclusion is well captured, see Figure 2.13(a)), whereas spurious oscillatory stresses are encountered at the interface with the standard Lagrange multiplier-based procedure due to mesh locking (see Figure 2.13(b)). To better appreciate the phenomenon, the jump of the traction forces across Γ with respect to the parametric coordinate along the interface is also plotted in Figure 2.14. These results confirm the attractiveness of the proposed non-invasive method over the standard approach: it can be easily extended with a suitable average operator to properly handle severe bi-material interfaces.

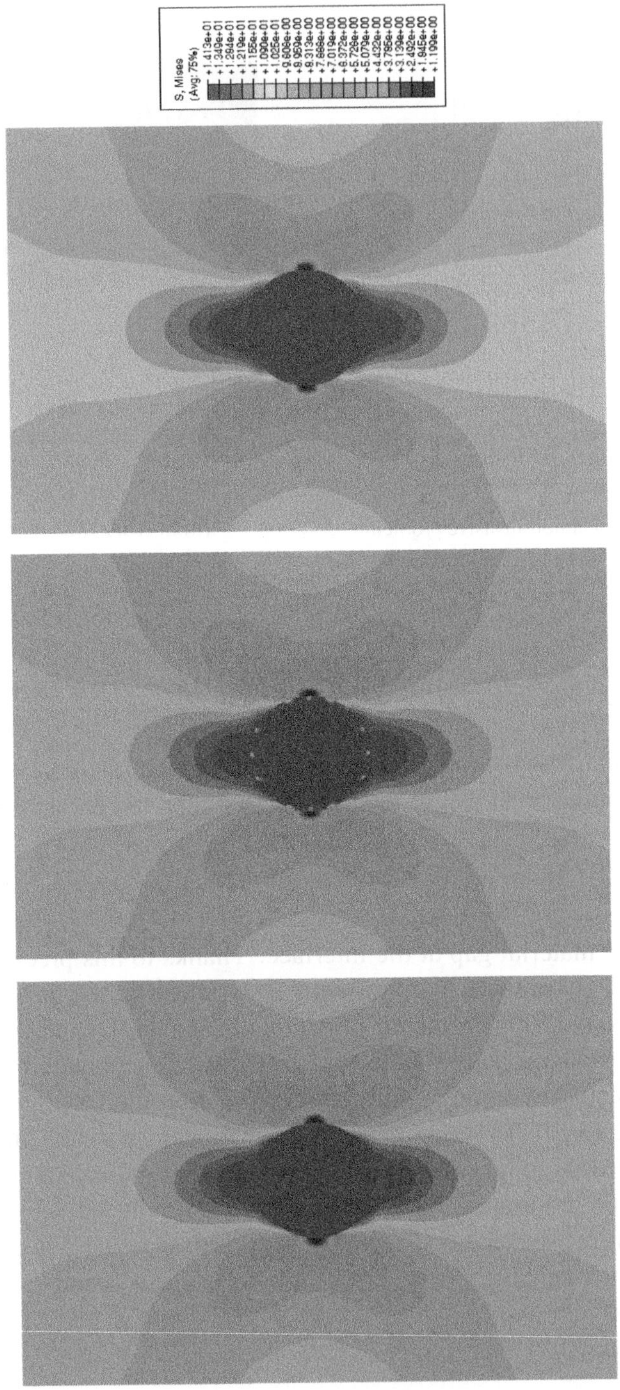

(a) Non-invasive Nitsche based strategy.

(b) Non-invasive Lagrange based strategy.

(c) Reference ABAQUS solution.

Figure 2.13. *Plate with a central inclusion: von Mises stress distribution. For a color version of this figure, see www.iste.co.uk/bouclier/IGA.zip*

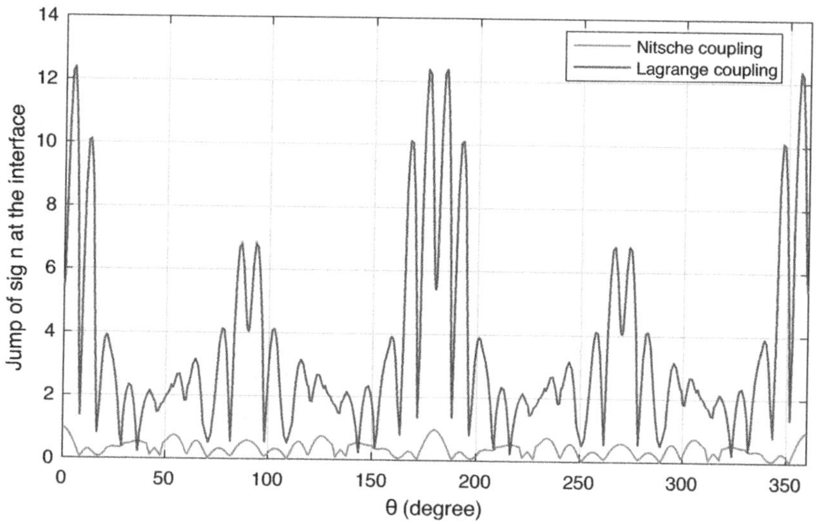

Figure 2.14. *Plate with a central inclusion: jump of the interface tractions along Γ (computed as $||\sigma_1 n_{11} + \sigma_2 n_2||_\Gamma$). For a color version of this figure, see www.iste.co.uk/bouclier/IGA.zip*

NOTE.– It should be noted here that the adopted modeling may not be the optimal one in terms of efficiency of the iterative strategy. Indeed, usually, the local region is chosen to be large enough so that the behavior around the interface Γ between the global and local models becomes closer (see, for example, Bouclier *et al.* (2016), which compares different modelings for such a test case).

NOTE.– Let us underline that the extension of Nitsche-based methods to material nonlinearities (e.g. plasticity or damage) does not have an obvious way forward. Indeed, the definition of the virtual mean interface resultant force is delicate in this context (see the second term in equation [1.44]). Nevertheless, following the philosophy of the previous note, what can be done in our case of global/local modeling is to take a local region that is large enough so that the local nonlinearity does not play a major role at the interface Γ. As a result, we end up with a coupling in a region where both models can be assumed to be linear elastic, thus making it possible to use the present approach without any additional ingredients.

2.5. Summary and discussion

The non-invasive global/local coupling approach has emerged in recent years in the field of standard FEM as a response to a present-day industrial issue: how to locally enrich a complex, costly global model without modifying it (and its associated numerical operators)? Beyond being an efficient methodology to couple different models, the method offers the possibility of combining different numerical codes with low implementation effort. From a mathematical point of view, the usual strategy starts with a Lagrange multiplier-based formulation of the resulting global/local coupling problem. Then, by making use of the additional interface unknowns, the obtained augmented system is split to make the global and local models communicate, in an iterative manner, through interface exchanges, while keeping the initial structure of their numerical operators.

Given the rigid tensor product structure of the multivariate (most commonly used) spline bases, the non-invasive global/local coupling approach is relevant to the field of IGA from a scientific point of view. It offers an original alternative to computing the resulting global/local trimmed isogeometric configuration. The idea is to take the initial isogeometric (possibly multi-) patch model as the global model, in order to eliminate the need for costly spline re-parameterization procedures when integrating a non-conforming local model. In addition, unlike standard unfitted coupling methods, the global stiffness operator is assembled and factorized only once, and the system to be solved remains well-conditioned, regardless of the evolution of the local model. However, more critically than in FEM, the challenge of handling a non-conforming coupling situation is crucial when extending the technology to IGA, since no simple meshing procedure can be achieved to recover the case of non-matching (or even matching) interfaces, as in FEM.

To address this issue, we propose to build a novel non-invasive global/local algorithm that makes use of the non-symmetric Nitsche coupling approach. Recently, the non-symmetric variant of the Nitsche method has successfully been applied to the weak enforcement of Dirichlet boundary and interface conditions in spline unfitted discretizations (Schillinger *et al.* 2016a; Guo 2017; Hu *et al.* 2018). The work in this chapter further consolidates this technique by establishing its application in the context of global/local isogeometric structural analysis. In this field, the non-symmetric Nitsche coupling gives the full potential of the non-invasive approach: it offers the opportunity of simply and efficiently addressing all the non-conforming global/local coupling configurations that may be encountered when integrating a local region within an underlying rigid global spline model. The reason for this is the robustness and simplicity of the coupling: it is (1) free of auxiliary fields, which enables us to circumvent the difficulty of choosing a suitable dual space (particularly in the case of non-conforming couplings), and (2) intrinsically stable (in a weak sense), thus eliminating the need for additional stabilization terms with appropriate parameters (whose determination can be delicate, depending on the cut scenarios). The obtained results clearly indicate the superiority of the developed algorithm when addressing any arbitrary non-conforming situations: we always achieve optimal accuracy, but the standard non-invasive Mortar-based version fails to reproduce optimal behavior in some configurations. In addition, mesh locking is reported when coupling at a severe bi-material interface with the standard Lagrange multiplier method, whereas the Nitsche counterpart handles such interfaces successfully, thanks to the use of a suitable average operator.

From an engineering point of view, such a technique may be of great interest since, in many applications, it is necessary to solve a sequence of similar problems in a multiresolution process, where only the behavior at a local scale needs to be updated (e.g. to carry out the shape optimization of local entities (Mulani *et al.* 2013; Fußeder *et al.* 2015; Hirschler *et al.* 2019a), to perform the identification of local mechanical parameters (Passieux *et al.* 2015; Réthoré 2015), or to model crack propagation (Passieux *et al.* 2013), or the expansion of a plastic zone (Gendre *et al.* 2009; Duval *et al.* 2016), or a damage zone (Guinard *et al.* 2018), etc.). In particular, since the objective of this book is to come up to the design of innovative structures using IGA, we will focus on the multiresolution occurring in shape optimization in Chapter 4. In this context, the proposed algorithm will deeply simplify the design modeling and ensure savings in computational time for the shape optimization of structural details within large macro-structures (see section 4.7).

Finally, regarding other applications of the developed method, we highlight once more that such a non-invasive procedure provides an iterative resolution of an unfitted problem while circumventing the conditioning issue. As a result, combining it with some recent FE non-invasive techniques that enable parallel computations while considering nonlinear constitutive laws (Duval *et al.* 2016; Oumaziz *et al.* 2018) may open the way for efficient iterative solution strategies in the general field of immersed boundary methods, which today represents a real challenge (de Prenter *et al.* 2017; Badia and Verdugo 2018; Jomo *et al.* 2019; Kerfriden *et al.* 2020).

3

Domain Decomposition Solvers for Efficient Multipatch IGA

3.1. Introduction

In this third chapter, we are interested in the efficient and accurate computation of globally complex (say industrial) structures modeled by IGA. The work presented here is fundamental for the final goal of this book, i.e. the design of innovative representative structures with improved properties. Although the developed methodology applies in a more general context, in the following, we pay special attention to tackling global aeronautic-type structures, i.e., broadly speaking, made up of complex curved stiffened panels (Totaro and Nicola 2012; Shroff *et al.* 2017; Guinard *et al.* 2018). These structures are obtained by assembling a main part (usually denoted by the skin or the panel) with subparts such as stringers and ribs (the stiffeners). The fastening of the stiffeners with the panel gives a final structure with a high stiffness-to-weight ratio. In this context, we are interested in having an independent discretization between the panel and the stiffeners to simplify the definition of the design space for the optimization (see Figure 1.21 and the corresponding discussion in section 1.4.3).

The use of IGA in this general field of computational solid mechanics, which combines (i) scientific computing, including shell analysis, and (ii) optimization (more precisely shape optimization), is highly relevant from a conceptual point of view, as demonstrated in the rich literature (see Lipton *et al.* (2010), Morganti *et al.* (2015) and Kamensky *et al.* (2017) for general simulations, Kiendl *et al.* (2009, 2015), Echter *et al.* (2013) and Maurin *et al.*

(2018) for Kirchhoff–Love shells, Benson *et al.* (2010), Dornisch *et al.* (2013), Echter *et al.* (2013) and Kikis *et al.* (2019) for Reissner–Mindlin shells, Bouclier *et al.* (2013b), Hosseini *et al.* (2013), Cardoso and Cesar de Sa (2014), Bouclier *et al.* (2015a, 2015b) and Caseiro *et al.* (2015) for solid-shells and, finally, Wall *et al.* (2008), Nagy *et al.* (2013), Kiendl *et al.* (2014), Fußeder *et al.* (2015), Bandara and Cirak (2017) and Hirschler *et al.* (2019b) for solid and shell shape optimization, to name a few). Now, truly addressing complex structures in this framework still requires us to face several challenges, the most crucial one probably being the non-conforming multipatch analysis issue since a single B-spline or NURBS patch cannot represent non-trivial geometries (see the previous discussion in section 1.4.2). This is also essential for bridging the gap between CAD (computer-aided design) and analysis with IGA. Focusing on shell structures, it is actually one of the two key issues (with trimming) to be addressed as the boundary representation (B-rep) of surfaces is directly available in CAD. In the particular case of shells, since we are interested in stiffened structures that are relatively thin, we will consider isogeometric Kirchhoff–Love shell elements (Kiendl *et al.* 2009; Echter *et al.* 2013), for which IGA is even more attractive because it naturally satisfies the C^1 requirement of such formulations. This enables us to derive a rotationless shell element, thereby minimizing the number of DOF (degree-of-freedom).

Multipatch isogeometric analysis generally goes with large-scale simulations. Efficient solvers for large-scale systems have been intensely studied over the past three decades by the scientific computing community. In structural mechanics, non-overlapping domain decomposition methods have shown strong potentials, and are now considered as robust strategies for solving complex problems. In this chapter, we propose to take advantage of these established strategies for performing multipatch isogeometric analysis. To this purpose, after motivating our choice of considering non-overlapping domain decomposition methods in this context, we first introduce a new Mortar coupling formulation for Kirchhoff–Love shells. Then, step by step, we build a family of dual domain decomposition algorithms that are suitable from any Mortar coupling formulation. Finally, we assess the performance of the developed solvers in the context of standard solid elasticity and of Kirchhoff–Love shells. In the same idea of what was performed in the previous chapter, the solver appears to us more appropriate in a multiresolution context, such as during shape optimization, since it allows us to factorize once and for all the stiffness matrices of patches located in non-design regions. From a conceptual point of view, a parallel can actually be drawn between this chapter and the previous one, as will be demonstrated in the following with several references to Chapter 2. The work presented in this chapter originates from Hirschler *et al.* (2019a, 2019c).

3.2. Benefiting from the additional Lagrange multiplier field for multipatch analysis

As seen previously in this book, various schemes based on weak couplings have been proposed over the last few years to connect arbitrary isogeometric domains, resulting in three main approaches: penalty coupling, Mortar coupling and Nitsche coupling (see especially section 1.5). Among all methods, various arguments are put forward regarding different criteria such as the simplicity, the accuracy, the robustness or even the range of applications. Nevertheless, a great potential of the Mortar technique does not seem to have been fully exploited in IGA yet: there is a close link between this class of coupling technologies and the family of domain decomposition (DD) algorithms.

In fact, the additional DOF associated with the Lagrange multiplier field of the Mortar approach should not be seen as a drawback (as is often the case in the IGA community) from a scientific computing point of view. Indeed, this additional field enables us to formulate an interface problem that is naturally suitable for parallel computing (Stefanica 2001, 2005). The point of interest thus lies in the potential of domain decomposition methods (DDM) for solving large-scale systems. These non-overlapping DDM have been actively studied over the past decades in FEM (finite element methods) as efficient parallel solvers for systems with possible millions of DOF (Gosselet and Rey 2006). The idea is to subdivide the computational domain into non-overlapping sub-domains: instead of solving one large system, multiple independent local systems are solved per sub-domain (in parallel) and interface conditions are recovered through an iterative process (see Figure 3.1). More specifically, the finite element tearing and interconnecting (FETI) (Farhat and Roux 1991) and the balancing domain decomposition (BDD) (Le Tallec *et al.* 1991; Mandel 1993) are often seen as originators of the two main branches of these non-overlapping DDM. BDD is usually referred to as a primal approach since it chooses the interface displacement field as the main unknown, whereas the FETI is usually referred to as a dual approach because it grants privileges to the interface reaction forces. Since then, a large number of papers have been published with numerous variants of these two methodologies that have been successfully applied in many contexts (e.g. in the field of solid mechanics, see Farhat *et al.* (1998, 2001) and Amini *et al.* (2009) for plates and shells, Jolivet *et al.* (2014), Gosselet *et al.* (2015) and Bovet *et al.* (2017) for heterogeneous problems, and Dureisseix and Farhat (2001) for contact problems).

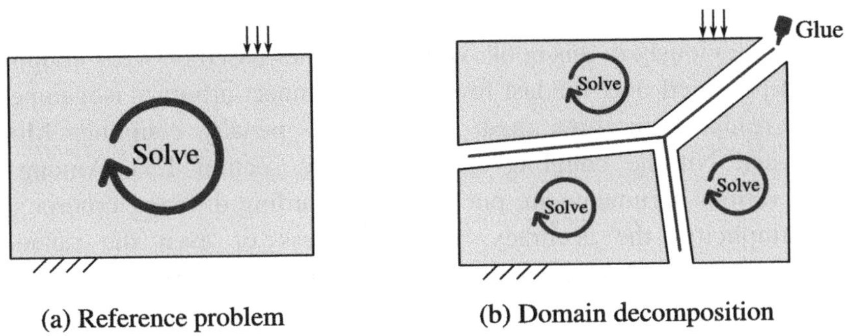

Figure 3.1. *Principle of non-overlapping domain decomposition methods (DDM): (a) instead of solving one large system defined over the whole computational domain* Ω*, (b) the domain is subdivided into multiple non-overlapping sub-domains* Ω_s*. The resolution is performed by solving, in parallel, local systems defined per sub-domain, and interface conditions are recovered through an iterative process. IGA offers a natural splitting of the initial domain into several sub-domains: each isogeometric patch can be associated with a sub-domain*

In the context of IGA, although marginally investigated, this class of solvers is all the more relevant as multipatch spline discretizations can be seen as a natural domain decomposition (each sub-domain being an isogeometric patch). Kleiss *et al.* (2012) noted this logical link between IGA and DDM, which led to the Isogeometric ETI (IETI) solver (Kleiss *et al.* 2012; Hofer and Langer 2017; Stavroulakis *et al.* 2018), which can be viewed as the direct isogeometric variant of the FETI dual–primal (FETI-DP) solver (Farhat *et al.* 2001) for the coupling of matching patches in standard solid elasticity. Based on all these discussions, we propose in the following to strongly enlarge the scope of DDM for IGA by addressing the computation of complex, non-conforming multipatch structures modeled using standard solid elasticity or Kirchhoff–Love shells. More precisely, we build dual DD algorithms that are efficient from a computational point of view (suitable for parallel computing) as well as accurate (quality of the resolution, especially at the interface junctions). In particular, when used in the context of shape optimization, such tools will become even more attractive since they will not allow us to recompute the parts of the structure in non-design areas at each optimization iteration (see section 4.3).

NOTE.– Regarding the coupling terminology, we obviously recycle the nomenclature introduced in section 1.4.2, i.e. we refer to it as a *matching interface* when it is aligned with some edges of the patches and the discretizations on both sides are exactly the same, as a *non-matching interface* when it is still aligned with some edges of the patches but the two discretizations differ, and as a *non-conforming interface* when it can cut some (knot-span) elements of the patches. In addition, it has to be said at this stage that we do not consider real trimmed IGA in the sense that some parts of the patches to be coupled are not removed from across the interface. Recalling that an emphasis is placed on stiffened structures in this chapter, what we actually characterize as a non-conforming coupling here is the situation of Figure 1.21 (bottom), i.e., roughly speaking, the coupling of a stiffener lying arbitrarily on a panel with a fixed discretization. Such a modeling will be sufficient to design advanced aeronautical structures (see Chapter 4 and more precisely section 4.6). The non-conforming coupling situation in this chapter thus slightly differs from that of the previous chapter where the global solution in pieces of elements was discarded (in favor of the solution of the local model, see Figure 2.1 as a reminder). Perhaps this partially explains why we are able to obtain good results with a Mortar-based approach for the coupling here (see sections 3.3.3 and 3.5).

3.3. Case of multipatch Kirchhoff–Love shell analysis

After these opening remarks, let us now focus on the coupling of isogeometric Kirchhoff–Love shell patches. Given the specific nature of the Kirchhoff–Love shell modeling, it is first necessary to develop a suitable coupling approach on which we will be able to build, in a second step, a DD algorithm. After recalling the basics of isogeometric Kirchhoff–Love shells, a new Mortar approach, which involves two sets of Lagrange multipliers in order to impose the continuity of the mid-surface displacement and of the rotation of the shell, is thus presented here and assessed through preliminary examples. Regarding the illustrations, an emphasis is placed on stiffened structures that are ubiquitous in aeronautics.

3.3.1. *Kirchhoff–Love shell formulation: basics*

3.3.1.1. *Kinematics*

The Kirchhoff–Love modeling has been largely studied in the literature for structural analysis and for shape optimization of thin structures. The starting point consists of invoking specific kinematic assumptions, namely the Kirchhoff–Love hypotheses. These hypotheses introduced by Kirchhoff (1850) and Love (1888) state that the normals to the mid-surface in the reference configuration remain normal and unstretched in the deformed configuration. In other words, the transversal deformations through the shell thickness are neglected in this model. It is commonly accepted that the Kirchhoff–Love model is often sufficient to describe, from a practical point of view, the structural behavior of thin-walled structures (Bischoff *et al.* 2017). Figure 3.2 presents these modeling choices.

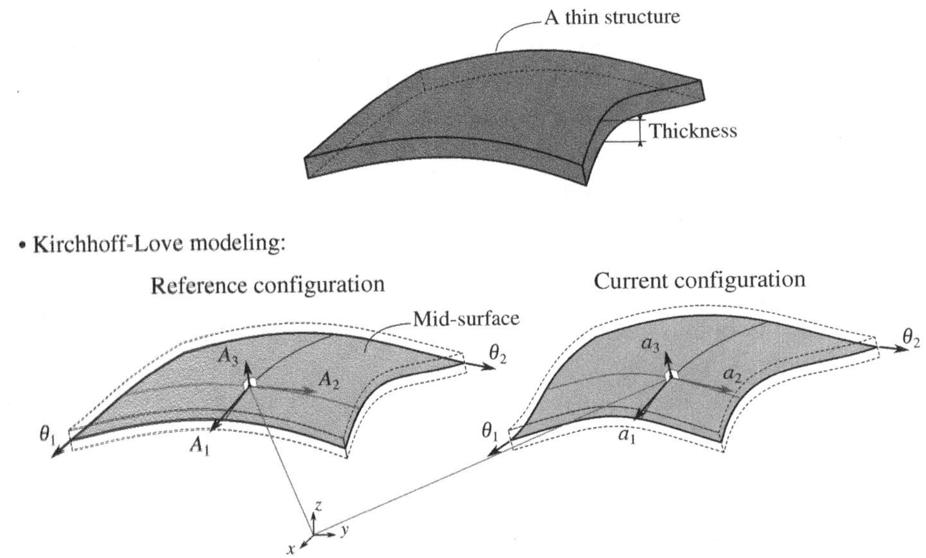

Figure 3.2. *Schematic view of a shell body and the associated Kirchhoff–Love modeling approach. For a color version of this figure, see www.iste.co.uk/bouclier/IGA.zip*

There are different strategies to introduce these kinematic assumptions in the continuum formulation. In this work, we follow the strategy from Kiendl *et al.* (2009) based on the direct approach. It means that the shell is regarded from the beginning as a surface (often called a Cosserat surface) and proper kinematic assumptions, representing the three-dimensional behavior, are postulated. More specifically, the three-dimensional shell body is described by its mid-surface and a director vector. To this end, two curvilinear coordinates θ_1 and θ_2 are introduced to uniquely identify a specific position on the mid-surface. In order to describe the out-of-plane dimension of the shell body, an additional parameter $\theta_3 \in [-t/2, t/2]$ is introduced, where t denotes the thickness of the shell. To summarize, the reference configuration of the shell is written as:

$$\Omega = \left\{X \in \mathbb{R}^3 : X = S(\theta_1, \theta_2) + \theta_3 A_3(\theta_1, \theta_2), \quad \forall(\theta_1, \theta_2, \theta_3) \in \bar{\Omega}\right\}, \tag{3.1}$$

where the parameter space is $\bar{\Omega} = \bar{\Omega}_0 \times [-t/2, t/2]$, with $\bar{\Omega}_0 \subset \mathbb{R}^2$ being the parameter space associated with the mid-surface S. The director vector A_3 is defined at each location on the mid-surface as the unit normal:

$$A_3 = \frac{1}{J} A_1 \times A_2, \quad \text{with} \quad J = |A_1 \times A_2|, \tag{3.2}$$

where A_1 and A_2 are the covariant basis vectors associated with the mid-surface:

$$A_\alpha = \frac{\partial S}{\partial \theta_\alpha} = S_{,\theta_\alpha} \quad \alpha = 1, 2. \tag{3.3}$$

The subscript $(\cdot)_{,\theta_\alpha} = \partial(\cdot)/\partial\theta_\alpha$ indicates the partial derivative with respect to the variable θ_α and $\times$ stands for the cross product between two vectors. Again, the reader is referred to Figure 3.2 where the aforementioned geometrical quantities are illustrated. Then, the deformed shell body is defined analogously to equation [3.1]. A material point, initially given by its reference position $X := X(\theta_1, \theta_2, \theta_3)$, is mapped after deformation to its current position $x := x(\theta_1, \theta_2, \theta_3)$. The current position vector is given by:

$$x(\theta_1, \theta_2, \theta_3) = s(\theta_1, \theta_2) + \theta_3 a_3(\theta_1, \theta_2), \tag{3.4}$$

where s denotes the deformed mid-surface and a_3 denotes its associated director vector. Finally, we introduce the displacement field of the mid-surface u such that the deformed mid-surface is given by:

$$s(\theta_1, \theta_2) = S(\theta_1, \theta_2) + u(\theta_1, \theta_2). \tag{3.5}$$

By recalling the Kirchhoff–Love assumptions, the director vector in the deformed configuration is taken as the unit vector to the deformed mid-surface as shown for the reference configuration (see equation [3.2]). It leads us to write the displacement field of the shell body as:

$$\begin{aligned} u^{\mathrm{KL}}(\theta_1,\theta_2,\theta_3) &= x(\theta_1,\theta_2,\theta_3) - X(\theta_1,\theta_2,\theta_3) \\ &= u(\theta_1,\theta_2) + \theta_3\,[a_3 - A_3]\,(\theta_1,\theta_2). \end{aligned} \quad [3.6]$$

The difference vector $w = a_3 - A_3$ is explicitly expressed as a function of the displacement field of the mid-surface (i.e. vector u). In the context of small displacements, the difference vector is approximated with its linearized form, which can be written as:

$$w^{\mathrm{lin}} = \Phi \times A_3\,, \quad [3.7]$$

where the (linearized) rotation vector Φ is defined by Echter *et al.* (2013):

$$\Phi = \phi_1 A_1 + \phi_2 A_2, \quad [3.8]$$

with the rotation angles ϕ_1 and ϕ_2 given by:

$$\phi_1 = \frac{1}{J} u_{,\theta_2} \cdot A_3 \quad \text{and} \quad \phi_2 = -\frac{1}{J} u_{,\theta_1} \cdot A_3. \quad [3.9]$$

To summarize, the displacement field of the whole body u^{KL} is described by quantities associated only with the mid-surface; namely the mid-surface displacement field u, and several geometrical quantities where the covariant vectors (A_1 and A_2) are the main vectors.

3.3.1.2. *Continuum mechanics*

Once the kinematics is set, the displacement field is introduced into a strain measure, and, for example, the Green–Lagrange strain tensor. Within the Kirchhoff–Love model, the deformations are usually considered as linear with respect to the thickness coordinates θ_3. The linearized Green–Lagrange strain tensor takes the following form:

$$\varepsilon^{\mathrm{KL}}(\theta_1,\theta_2,\theta_3) = \mathrm{e}(\theta_1,\theta_2) + \theta_3 \kappa(\theta_1,\theta_2) \quad [3.10]$$

where e and κ are the membrane and bending strain tensor, respectively. Their strain components expressed in the contravariant basis are given by:

$$\mathrm{e}_{\alpha\beta} = \frac{1}{2}\Big(u_{,\theta_\alpha} \cdot A_\beta + u_{,\theta_\beta} \cdot A_\alpha\Big), \quad [3.11]$$

and

$$\kappa_{\alpha\beta} = -u_{,\theta_\alpha\theta_\beta} \cdot A_3 + \frac{1}{J}\Big[u_{,\theta_1} \cdot \big(A_{\alpha,\theta_\beta} \times A_2\big) + u_{,\theta_2} \cdot \big(A_1 \times A_{\alpha,\theta_\beta}\big)\Big]$$

[3.12]

$$+ \frac{A_3 \cdot A_{\alpha,\theta_\beta}}{J}\Big[u_{,\theta_1} \cdot \big(A_2 \times A_3\big) + u_{,\theta_2} \cdot \big(A_3 \times A_1\big)\Big].$$

The Greek subscripts take the values 1 and 2 (as stated above, the transverse shear strains vanish, i.e. $\varepsilon_{i\alpha} = 0, \ \forall \alpha$).

When deriving shell theories by the direct approach, two-dimensional material equations are postulated (Bischoff *et al.* 2017). Under the assumptions of a linear elastic and isotropic behavior of the material, the stress–strain relation is defined by the constitutive fourth-order tensor C_0 where the contravariant components are:

$$C_0^{\alpha\beta\gamma\delta} = \bar{\lambda} A^{\alpha\beta} A^{\gamma\delta} + \mu \left(A^{\alpha\gamma} A^{\beta\delta} + A^{\alpha\delta} A^{\beta\gamma} \right), \qquad [3.13]$$

and $\bar{\lambda} = 2\lambda\mu/(\lambda + 2\mu)$, with λ and μ being the Lamé constants. $A^{\alpha\beta}$ denotes the mid-surface contravariant metrics. This constitutive law includes the plane stress condition by the condensation of the material equations (which leads to the parameter $\bar{\lambda}$). For more information on the expressions of the curvilinear quantities (covariant and contravariant vectors, metrics, etc.), the reader may refer to Kiendl (2011) and Bischoff *et al.* (2017).

Finally, the virtual works for the Kirchhoff–Love shell formulation are given by:

$$\delta \mathrm{W}_{\mathrm{int}}^{\mathrm{KL}} = - \int_{\bar{\Omega}_0} \left(\mathrm{n} : \delta \mathrm{e} + \mathrm{m} : \delta \kappa \right) J \, \mathrm{d}\bar{\Omega}_0 \,, \qquad [3.14]$$

$$\delta \mathrm{W}_{\mathrm{ext}}^{\mathrm{KL}} = \int_{\bar{\Omega}_0} \delta u \cdot \mathrm{f}_s \, J \, \mathrm{d}\bar{\Omega}_0 + \int_{\Gamma_0} \delta u \cdot \mathrm{f}_e \, \mathrm{d}\Gamma_0 \,. \qquad [3.15]$$

where the vector f_s describes distributed loads per unit of area applied on the mid-surface, and the vector f_e describes axial forces per unit of length applied on the edges of the mid-surface. The internal work is expressed as the sum of two contributions: the membrane and the bending part. Let us point out that the integration over the shell thickness (i.e. over parameter θ_3) has been performed analytically in equations [3.14] and [3.15]. This integration step introduces the normal forces n and the bending moments m. They are expressed as:

$$\mathrm{n} = t\, C_0 : \mathrm{e}, \tag{3.16}$$

$$\mathrm{m} = \frac{t^3}{12} C_0 : \kappa. \tag{3.17}$$

The interested reader can find further information about the Kirchhoff–Love shell modeling, for example, in Kiendl (2011) and Bischoff *et al.* (2017).

3.3.1.3. *Element formulation*

An isogeometric Kirchhoff–Love shell element can be built by discretizing the mid-surface with a bivariate spline. As for standard IGA solid elements, the displacement field is then discretized using the same basis functions as for the mid-surface. However, there are some restrictions on the degree and the smoothness of this discretization. In fact, the bending terms involve second-order derivatives, as seen in equation [3.12]. Therefore, splines of degree two or higher should be used. Moreover, the chosen discretization has to be C^1-continuous such that the bending behavior of the shell is well described. If the model presents C^0-continuity at some location, then a specific treatment has to be done there (Kiendl *et al.* 2010).

Finally, the equilibrium configuration of the shell followed from the principle of minimum potential energy leads to the typical linear system:

$$\mathbf{K}\mathbf{u} = \mathbf{f}, \tag{3.18}$$

where the vector $\mathbf{u}$ concatenates the DOF of the mid-surface displacement field, and $\mathbf{K}$ and $\mathbf{f}$ denote the isogeometric Kirchhoff–Love stiffness and external force operators, respectively. These operators are obtained by introducing the discretized quantities into the internal [3.14] and external work [3.15]. Finally, these operators can be computed as standard solid elements: numerical quadrature rules are used at the element level to build elementary operators and an assembling procedure iteratively adds these local contributions to the global operators. For example, the stiffness matrix of this Kirchhoff–Love shell formulation can be built through 3-by-3 matrices of the form:

$$\mathbf{K}^e_{kl} = \int_{\bar{\Omega}^e_0} \left[t\mathbf{B}^{m\,T}_k \mathbf{H}_0 \mathbf{B}^m_l + \frac{t^3}{12} \mathbf{B}^{b\,T}_k \mathbf{H}_0 \mathbf{B}^b_l \right] J \mathrm{d}\theta_1 \mathrm{d}\theta_2, \tag{3.19}$$

where k and l are indices of two control points related to the element e of domain $\bar{\Omega}_0^e$ in the mid-surface parameter space. The expression of the membrane strain matrices $\mathbf{B}_k^m$ and the bending strain matrices $\mathbf{B}_k^b$ can be inferred from equations [3.11] and [3.12]. The matrix $\mathbf{H}_0$ represents the constitutive tensor [3.13] following the Voigt notation. Further details on the expression of this isogeometric Kirchhoff–Love shell element can be found in Kiendl (2011), and, for example, in Cirak *et al.* (2000), Kiendl *et al.* (2015) and Hirschler *et al.* (2019b).

3.3.2. *Formulation of the coupled problem*

The general field of shell coupling has been subjected to intensive studies these last five years. In this context, all of the three main methods have been investigated (see, for example, Apostolatos *et al.* (2014a), Breitenberger *et al.* (2015), Horger *et al.* (2019), Herrema *et al.* (2019), Leidinger *et al.* (2019), Coradello *et al.* (2020) and Leonetti *et al.* (2020) for penalty coupling, Apostolatos *et al.* (2014a), Dornisch *et al.* (2017), Duong *et al.* (2017), Coox *et al.* (2017b), Sommerwerk *et al.* (2017), Hirschler *et al.* (2019a, 2019c), Schuß *et al.* (2019), and Paul *et al.* (2020) for Mortar coupling, and Guo and Ruess (2015), Guo (2017), Nguyen-Thanh *et al.* (2017), Guo *et al.* (2018) and Liu *et al.* (2019) for Nitsche coupling). Focusing on Kirchhoff–Love shells, all three techniques have been successfully applied for large and complex structures, even for nonlinear analysis (Apostolatos *et al.* 2014a; Breitenberger *et al.* 2015; Duong *et al.* 2017; Nguyen-Thanh *et al.* 2017; Guo *et al.* 2018; Herrema *et al.* 2019; Schuß *et al.* 2019; Leonetti *et al.* 2020; Paul *et al.* 2020). Finally, regarding the specific case of Mortar coupling for Kirchhoff–Love shells, it appears that most of the studies are restricted to the connection of G^1 surfaces (Apostolatos *et al.* 2014a; Schuß *et al.* 2019). In this work, we seek to build a more general formulation that can handle multipatch Kirchhoff–Love shells which connect with arbitrary angles in the context of linear elasticity. Indeed, recalling that one of our intended applications is the computation of stiffened structures, the junctions between the panel and the stiffeners have only C^0 regularity along the non-conforming interfaces (see Figure 1.21 (bottom)). Some aspects can be found in Duong *et al.* (2017), Paul *et al.* (2020) and Leonetti *et al.* (2020), but the applicability in the context of linear analysis may not be obvious: the kinematic constraints are built by dot and cross products between quantities of neighboring patches, which result in nonlinear terms (with respect to displacements). It thus requires a linearization and the resolution is performed by the Newton–Raphson method (see Duong *et al.* (2017), appendix D.2). The main differences with our approach lie in the kinematic constraints imposed by the Lagrange multiplier. Similarities with already published papers in the standard FEM community can be found with our method, especially with Bernadou *et al.* (1989) and Bernadou and Cubier (1998) where theoretical studies have been carried out.

3.3.2.1. *Kinematic coupling conditions*

For the sake of simplicity, the method is presented in the case of two non-overlapping sub-domains Ω_1 and Ω_2, without loss of generality. Subscripts 1 and 2 are now used to denote quantities related to sub-domains Ω_1 and Ω_2, respectively. At the common interface denoted by Γ, the continuity of the displacement has to be clearly ensured. Furthermore, in the case of a shell junction with a rigid hinge, an additional constraint is required, which enforces the continuity of the rotation in the tangential direction associated with the interface curve (see Figure 3.3). Thus, we formulate the kinematic constraints as:

$$u_1 = u_2 \quad \text{on } \Gamma \qquad [\text{3 displacements}], \qquad [3.20]$$

$$\Phi_1 \cdot A_\Gamma = \Phi_2 \cdot A_\Gamma \quad \text{on } \Gamma \qquad [\text{1 rotation}], \qquad [3.21]$$

where A_Γ is a unit tangent vector associated with the interface curve.

3.3.2.2. *Weak coupling with a Mortar approach*

The two sets of constraints are ensured in a weak sense by introducing two Lagrange multipliers $\lambda_d \in \mathcal{M}_d$ and $\lambda_r \in \mathcal{M}_r$ ($\mathcal{M}_d$ and $\mathcal{M}_r$ being *ad hoc* spaces). It results in the formulation of the following Lagrangian for the coupled problem:

$$\begin{aligned} L_{\text{KL}}(u_1, u_2, \lambda_d, \lambda_r) = &\ \tfrac{1}{2} a_1(u_1, u_1) - l_1(u_1) \\ &+ \tfrac{1}{2} a_2(u_2, u_2) - l_2(u_2) \\ &+ b(\lambda_d, u_1 - u_2) \\ &+ b(\lambda_r, \Phi_1 \cdot A_\Gamma - \Phi_2 \cdot A_\Gamma), \end{aligned} \qquad [3.22]$$

where bilinear forms a_s and linear forms l_s constitute the standard variational forms of the Kirchhoff–Love shell problem on each sub-domain ($s \in \{1, 2\}$), and b is defined as in equation [1.42] (see section 1.5.3).

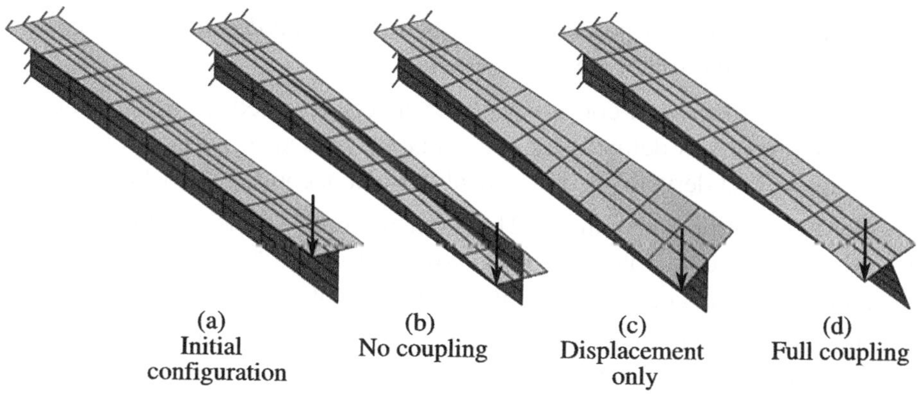

Figure 3.3. *Kinematic constraints for Kirchhoff–Love shells: at the patch junction, obviously, the three components of the displacement fields need to be coupled (c). In addition, the continuity of the rotation in the tangent direction of the interface curve has to be imposed (d)*

Then, repeating the procedure of the previous chapters (see sections 1.5.3 and 2.2.2), the variational principle written in the discrete form gives the following coupled linear system to be solved:

$$\begin{bmatrix} \mathbf{K}_1 & \mathbf{0} & \mathbf{C}_1^T & \mathbf{F}_1^T \\ \mathbf{0} & \mathbf{K}_2 & -\mathbf{C}_2^T & -\mathbf{F}_2^T \\ \mathbf{C}_1 & -\mathbf{C}_2 & \mathbf{0} & \mathbf{0} \\ \mathbf{F}_1 & -\mathbf{F}_2 & \mathbf{0} & \mathbf{0} \end{bmatrix} \begin{pmatrix} \mathbf{u}_1 \\ \mathbf{u}_2 \\ \boldsymbol{\lambda}_d \\ \boldsymbol{\lambda}_r \end{pmatrix} = \begin{pmatrix} \mathbf{f}_1 \\ \mathbf{f}_2 \\ \mathbf{0} \\ \mathbf{0} \end{pmatrix}, \qquad [3.23]$$

where vectors $\mathbf{u}_1$, $\mathbf{u}_2$, $\boldsymbol{\lambda}_d$ and $\boldsymbol{\lambda}_r$ represent the DOF corresponding to the discretizations of u_1, u_2, λ_d and λ_r, respectively. As in the previous chapter, the Mortar operators $\mathbf{C}_s$ and $\mathbf{F}_s$ consist of sparse rectangular matrices (see Hirschler *et al.* (2019c) for more information on their construction). Again and for the rest of this chapter, note that we use capital letters for matrices and lowercase letters for vectors to ease the understanding with respect to the implementation.

As for any mortar method, special care is required for the construction of the approximation subspaces of the Lagrange multipliers (see section 1.5.3).

Let p denote the smaller polynomial degree of both sub-domain displacement fields. We adopt the following strategy here:

– for the displacement constraint [3.20], a vector-valued spline function λ_d^h with degree $p - 1$ is defined since it is mainly related to normal forces (i.e. involving the first derivatives of the displacement according to [3.11]);

– for the rotation constraint [3.21], a scalar-valued spline function λ_r^h with degree $p - 2$ is defined because it is associated with a bending moment (i.e. involving the second derivatives of the displacement according to [3.12]);

– the same mesh refinement is chosen for both Lagrange multipliers λ_d^h and λ_r^h. We discretize these fields using a coarse (univariate) mesh with about as many elements as the coarsest of the sub-domains Ω_1 and Ω_2 over the interface.

NOTE.– As in the previous chapter, it is important to say that the choice for the dual discrete fields is based on physical considerations and numerical experiments. No rigorous proofs will be given in this book. However, let us emphasize that we make sure to take a mesh for the Lagrange multipliers (at least) that is as coarse as the coarsest of the sub-domains to be coupled. This is consistent with the theory (see, for example, Brivadis *et al.* (2015) and Schuß *et al.* (2019)), and certainly prevents us from obtaining overconstrained solutions. We assert that we never encountered spurious, undesirable phenomena in our computations. All numerical examples presented in this chapter and later in Chapter 4 highlight the good behavior of this Mortar coupling to handle non-conforming shell interfaces with arbitrary angles, as is the case, for example, with stiffened structures. Moreover, we can also note that the choice of the approximation spaces for the Lagrange multipliers is completely decoupled from the overall approach applied in this chapter to build the DD solver (see section 3.4). With other choices to hand, the remaining part of the method performs identically.

3.3.3. *Preliminary results: monolithic resolution*

Before the development of a dual DD algorithm that allows for an efficient resolution in the case of numerous patches, we present here preliminary numerical examples, where analytical results are known, in order to highlight the correctness of our Mortar coupling for Kirchhoff–Love shells. The resolution is performed with a direct solver applied to the coupled linear system [3.23].

3.3.3.1. *Bending and shear of a simple beam*

We begin by investigating a cantilever beam subjected to bending and shear due to a load applied on its right end (see Figure 3.4 (left)). Using the Euler–Bernoulli beam theory, it is possible to analytically express the vertical displacement, shear force and bending moment. We adopt the following parameters: beam length $L = 20$ m, width $w = 1$ m, thickness $t = 0.1$ m, line load $q = 100$ Nm^{-1}, Young's modulus $E = 60$ GPa and Poisson's ratio $\nu = 0$. Note that we used a Poisson ratio $\nu = 0$ so as to be consistent with the beam theory. The beam is decomposed into two non-matching cubic patches of the same size (i.e. the interface is located at half the length), as shown in Figure 3.4 (left). Two shell elements are considered for the left part, while one shell element is used for the right part. Applying the strategy described above for the coupling, the Lagrange multipliers are discretized using the following (univariate) mesh: one quadratic B-spline element for the displacement constraint, and one linear B-spline element for the rotation constraint.

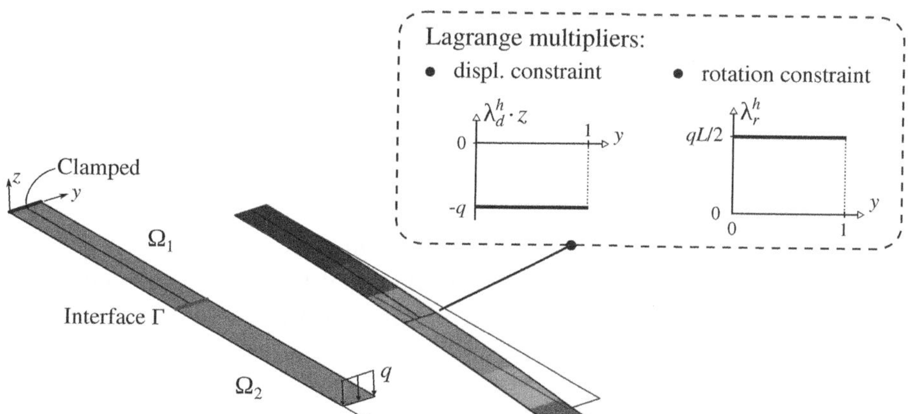

Figure 3.4. *Cantilever beam with end load: the beam is decomposed into two non-matching patches and coupled with the proposed Mortar approach. The shape of the Lagrange multipliers for the displacement and the rotation continuity constraints are shown on the right. For a color version of this figure, see www.iste.co.uk/bouclier/IGA.zip*

We run the analysis and verify that we exactly obtain the theoretical results since the analytical deflection is expressed as a polynomial of degree 3. More interestingly, the values obtained for the Lagrange multiplier fields λ_d^h and λ_r^h are:

$$\lambda_d^h \cdot z = -q = -100 \text{ Nm}^{-1} \qquad \text{and} \qquad \lambda_r^h = \frac{qL}{2} = 1000 \text{ N}, \quad [3.24]$$

which exactly correspond to the analytical values of the shear force and bending moment, respectively (note that we should consider the patch Ω_2 as the master, and the patch Ω_1 as the slave in order to obtain the same sign as in equation [3.24]). This test case demonstrates the mechanical meaning of the proposed Lagrange multiplier formulation. It is also useful for validating the implementation of the coupling operators. Finally, instead of putting the interface midway, we can cut the beam at different locations and investigate whether the analytical expressions are recovered.

3.3.3.2. *T-shape beam*

The T-beam problem is built using two planar patches that are connected in a non-conforming way, as shown in Figure 3.5(a). The interface crosses the middle elements of the upper patch. The upper patch has one more element than the other in the beam direction, and both patches are discretized with cubic elements. We take the same numerical setting as in Herrema *et al.* (2019), which is also given in Figure 3.5. We obtain a similar deformation as in Herrema *et al.* (2019) where penalty coupling was performed. In order to show that the rotation constraint is well prescribed, we plot in Figure 3.5(c) the resulting angle between the patches along the interface. It can be seen that it remains equal to 90° after deformation. Without the rotation constraint, the angle between the patches was found to be equal to 86.6° at the end of the beam because the interface acted as a hinge (Herrema *et al.* 2019). The presence of small oscillations for the coarser parameterization may be due to the non-optimality of the chosen spaces for the Lagrangian fields. Once again, it would be interesting to investigate the *inf–sup* condition to clarify the stability of the method (Wunderlich *et al.* 2019), but this is beyond the scope of this book. However, these oscillations vanish with a finer parameterization which, in our opinion, is satisfying.

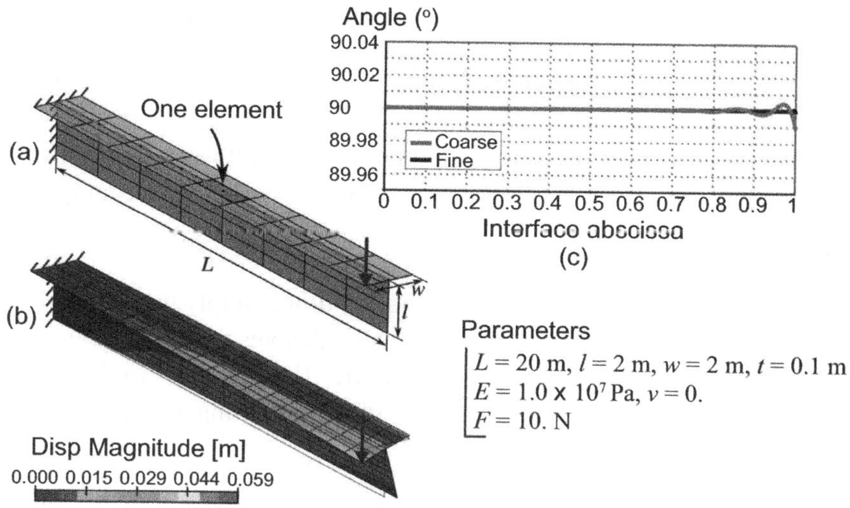

Figure 3.5. *T-Beam non-conforming coupling. (a) Description of the problem: the interface cuts the upper patch at the middle of some elements. (b) Deformed configuration with a scale factor of 10 and the magnitude of the displacement field. (c) Angle between the patches along the interface after deformation. The coarse case corresponds to the mesh shown in (b). The fine mesh is refined three times. For a color version of this figure, see www.iste.co.uk/bouclier/IGA.zip*

3.4. On the construction of dual domain decomposition solvers

Henceforth, we address the general case of n_s sub-domains (i.e. n_s patches) Ω_s, $s = 1 \ldots n_s$. The development proposed in this section is very general in the sense that it can be applied from any Mortar coupling formulation. Indeed, the family of DD algorithms that we build here start with a discrete system, which can be formally written as:

$$\begin{bmatrix} \mathbf{K} & \mathbf{C}^{F^T} \\ \mathbf{C}^F & \mathbf{0} \end{bmatrix} \begin{pmatrix} \mathbf{u} \\ \boldsymbol{\lambda} \end{pmatrix} = \begin{pmatrix} \mathbf{f} \\ \mathbf{0} \end{pmatrix}. \qquad [3.25]$$

This simply constitutes the generalization to several sub-domains of any coupled system using the Mortar approach. In this work, the constructed DD algorithms will be implemented in the context of standard solid elasticity and Kirchhoff–Love shells. Therefore, more precisely, we begin with the coupled systems of form [2.1] and [3.23], respectively. To arrive at system [3.25], we define as $\mathbf{K} = \mathrm{diag}(\mathbf{K}_1, \mathbf{K}_2, \ldots, \mathbf{K}_{n_s})$ the block diagonal matrix that stores all the stiffness matrices of all the sub-domains. We also define the resulting load vector $\mathbf{f} = (\mathbf{f}_1^T, \mathbf{f}_2^T, \dots, \mathbf{f}_{n_s}^T)^T$. The total displacement and Lagrange multiplier DOF are stored in the column vectors $\mathbf{u}$ and $\boldsymbol{\lambda}$, respectively. In the case of several Lagrange multiplier fields, we also concatenate the different Mortar coupling matrices into single coupling matrices $\mathbf{C}_s^F$ such that using our Mortar coupling approach for Kirchhoff–Love shells leads to, for example:

$$\mathbf{C}_s^F = \left[\mathbf{C}_s^T, \mathbf{F}_s^T\right]^T. \qquad [3.26]$$

In the case of a single Lagrange multiplier field, we simply have $\mathbf{C}_s^F = \mathbf{C}_s$. The matrices $\mathbf{C}_s^F$ are of size $n_\lambda \times n_{u_s}$, where n_λ denotes the size of the whole $\boldsymbol{\lambda}$, while n_{u_s} denotes the size of the local $\mathbf{u}_s$. Finally, the last ingredient to end up with [3.25] is to introduce the global coupling operator $\mathbf{C}^F = [\mathbf{C}_1^F, -\mathbf{C}_2^F, \dots, \mathbf{C}_{n_s}^F]$.

The aim of this section is to develop an efficient algorithm to solve a coupled linear system of form [3.25] based on the DDM, as in section 3.2. More precisely, we build a family of isogeometric, non-conforming, dual DD solvers by drawing inspiration from the initial one-level FETI approach (Farhat and Roux 1991).

NOTE.– For the specific situation of fourth-order problems as is the case with the Kirchhoff–Love shell theory, for example, the FETI-DP (see Farhat *et al.* (1998, 2001)) (and BDD by constraints (BDDC), see Amini *et al.* (2009)) has been proposed in the classical FEM framework. However, in the case of non-conforming interfaces, this approach seems to be difficult to apply: it is less trivial to identify the equivalent of the so-called corner nodes (subset of global DOF where exact continuity is enforced) since non-conforming parameterizations do not share common DOF by definition. This is why, in this book, we resort to the initial one-level FETI approach (Farhat and Roux 1991), even in the case of Kirchhoff–Love shells.

3.4.1. *Formulation of the interface problem*

The first step of the approach consists of splitting the problem [3.25] into the following coupled set of equations:

$$\mathbf{K}_s\mathbf{u}_s = \mathbf{f}_s - \mathbf{C}_s^{F^T}\boldsymbol{\lambda}, \quad \text{for } s = 1, \dots, n_s \qquad [3.27]$$

$$\sum_{s=1}^{n_s} \mathbf{C}_s^F \mathbf{u}_s = \mathbf{0}. \qquad [3.28]$$

The goal is then to formulate a dual interface problem with the Lagrange multiplier DOF vector as the only unknown. This interface problem is obtained by introducing the local equilibria [3.27] into the coupling condition [3.28].

3.4.1.1. *Equilibrium of floating sub-domains*

Generally, the local stiffness matrices $\mathbf{K}_s$ are not invertible. Indeed, if no Dirichlet boundary conditions are imposed on the floating sub domain Ω_s, then the corresponding matrix $\mathbf{K}_s$ is singular. Therefore, $\mathbf{K}_s$ being symmetric, equation [3.27] has a solution if and only if the right-hand side $\mathbf{f}_s - \mathbf{C}_s^{F^T}\boldsymbol{\lambda}$ belongs to the image of the operator $\mathbf{K}_s$. It leads to an additional equation, called the admissibility condition, which is:

$$\mathbf{R}_s^T\left(\mathbf{f}_s - \mathbf{C}_s^{F^T}\boldsymbol{\lambda}\right) = \mathbf{0}, \quad \text{for } s = 1, \ldots, n_s, \qquad [3.29]$$

where $\mathbf{R}_s$ is a rectangular matrix. Its columns describe a basis of the null space of the matrix $\mathbf{K}_s$, denoted by $\ker(\mathbf{K}_s)$. From a mechanical point of view, these vectors are the local rigid-body modes. Then, for each sub-domain, the equilibrium is given by:

$$\mathbf{u}_s = \mathbf{K}_s^+\left(\mathbf{f}_s - \mathbf{C}_s^{F^T}\boldsymbol{\lambda}\right) + \mathbf{R}_s\boldsymbol{\alpha}_s, \qquad [3.30]$$

where $\boldsymbol{\alpha}_s$ is a vector that contains the different amplitudes of the rigid-body modes (or null-energy modes) for the sub-domain s and $\mathbf{K}_s^+$ is a pseudo-inverse of the stiffness matrix, i.e. it satisfies $\mathbf{K}_s\mathbf{K}_s^+\mathbf{K}_s = \mathbf{K}_s$. If the stiffness matrix is non-singular, then the pseudo-inverse is unique and is equal to the inverse of the matrix $\mathbf{K}_s$, i.e. $\mathbf{K}_s^+ = \mathbf{K}_s^{-1}$.

3.4.1.2. *Substitution*

Now, the local solutions [3.30] can be introduced into the compatibility condition [3.28]. It yields the following equation:

$$\sum_{s=1}^{n_s}\mathbf{C}_s^F\mathbf{K}_s^+\mathbf{C}_s^{F^T}\boldsymbol{\lambda} - \sum_{s=1}^{n_s}\mathbf{C}_s^F\mathbf{R}_s\boldsymbol{\alpha}_s = \sum_{s=1}^{n_s}\mathbf{C}_s^F\mathbf{K}_s^+\mathbf{f}_s. \qquad [3.31]$$

Finally, combining this last equation with the admissibility condition given by [3.29], we end up with the interface problem. The resulting system is generally called the FETI system, which is given by:

$$\begin{bmatrix}\mathbf{F} & \mathbf{G}\\ \mathbf{G}^T & \mathbf{0}\end{bmatrix}\begin{pmatrix}\boldsymbol{\lambda}\\ \boldsymbol{\alpha}\end{pmatrix} = \begin{pmatrix}\mathbf{d}\\ \mathbf{e}\end{pmatrix}, \qquad [3.32]$$

where $\boldsymbol{\alpha}$ concatenates the rigid-body unknowns of each sub-domain. The dual Schur complement $\mathbf{F}$, the constraint matrix $\mathbf{G}$ and the right-hand sides $\mathbf{d}$ and $\mathbf{e}$ are expressed as:

$$\mathbf{F} = \sum_{s=1}^{n_s} \mathbf{C}_s^F \mathbf{K}_s^+ \mathbf{C}_s^{F^T}, \quad [3.33]$$

$$\mathbf{G} = -\begin{bmatrix} \mathbf{C}_1^F \mathbf{R}_1 & \mathbf{C}_2^F \mathbf{R}_2 & \cdots & \mathbf{C}_{n_s}^F \mathbf{R}_{n_s} \end{bmatrix}, \quad [3.34]$$

$$\mathbf{d} = \sum_{s=1}^{n_s} \mathbf{C}_s^F \mathbf{K}_s^+ \mathbf{f}_s, \quad [3.35]$$

$$\mathbf{e} = -\begin{pmatrix} \mathbf{f}_1^T \mathbf{R}_1 & \mathbf{f}_2^T \mathbf{R}_2 & \cdots & \mathbf{f}_{n_s}^T \mathbf{R}_{n_s} \end{pmatrix}^T. \quad [3.36]$$

3.4.2. *Solving the interface problem*

The resolution of system [3.32] relies on an iterative solver where only matrix–vector products are performed. Therefore, there is no need for assembling the dual Schur complement, which would be computationally very demanding since pseudo-inverses $\mathbf{K}_s^+$ are involved. Moreover, it avoids inverting the dual Schur complement that has a dense structure. More precisely, we call upon a conjugate projected gradient (CPG)-type algorithm. The idea is to iteratively solve a symmetric positive definite linear system (involving $\mathbf{F}$) under the constraint [3.29]. In the following, we give some details about the algorithm. Finally, note that we will improve the algorithm by building an adapted preconditioner (see section 3.4.4), thereby ending up with an efficient preconditioned-CPG (PCPG) solver to analyze isogeometric multipatch models.

3.4.2.1. *Projection*

From a global point of view, starting with the well-known CG algorithm, an additional projection step enables us to eliminate the DOF vector associated with the rigid-body motions of the floating sub-domains. A projector $\mathbf{P}$ onto the null space of $\mathbf{G}^T$, i.e. onto $\ker(\mathbf{G}^T)$, is introduced to satisfy the admissibility of the loading at every iteration of the resolution (i.e. equation [3.29] is satisfied for each sub-domain). By definition, such a projector verifies the relations:

$$\mathbf{G}^T\mathbf{P} = \mathbf{0}, \quad \text{and} \quad \mathbf{P}^T\mathbf{G} = \mathbf{0}.$$

It allows us to eliminate the unknowns $\boldsymbol{\alpha}$ from the first set of equations of system [3.32]. To this end, we multiply these equations on the left by $\mathbf{P}^T$:

$$\mathbf{P}^T\Big[\quad \mathbf{F}\boldsymbol{\lambda} + \mathbf{G}\boldsymbol{\alpha} = \mathbf{d} \quad\Big] \quad \Rightarrow \quad \mathbf{P}^T\mathbf{F}\boldsymbol{\lambda} = \mathbf{P}^T\mathbf{d}. \qquad [3.37]$$

The second equation $\mathbf{G}^T\boldsymbol{\lambda} = \mathbf{e}$ of system [3.32] is explicitly fulfilled by introducing the splitting:

$$\boldsymbol{\lambda} = \boldsymbol{\lambda}^{(0)} + \mathbf{P}\bar{\boldsymbol{\lambda}}, \qquad [3.38]$$

where the initialization $\boldsymbol{\lambda}^{(0)}$ satisfies the constraint $\mathbf{G}^T\boldsymbol{\lambda}^{(0)} = \mathbf{e}$. This splitting is operated during the resolution by first choosing $\boldsymbol{\lambda}^{(0)}$ as the starting guess, and second by projecting the gradient during the update. Finally, these two projection steps transform the interface problem [3.32] into the projected interface problem:

$$\mathbf{P}^T\mathbf{F}\mathbf{P}\,\bar{\boldsymbol{\lambda}} = \mathbf{P}^T\left(\mathbf{d} - \mathbf{F}\boldsymbol{\lambda}^{(0)}\right), \qquad [3.39]$$

which is solved using the PCPG algorithm, as described in Table 3.1.

We still need to express the projector and the initialization. This brings into play a symmetric matrix denoted by $\mathbf{Q}$ for which the product $\mathbf{G}^T\mathbf{Q}\mathbf{G}$ is invertible. The matrix $\mathbf{Q}$ can be taken as being the preconditioner, identity or a scaling matrix (Rixen *et al.* 1999). Finally, we have:

$$\boldsymbol{\lambda}^{(0)} = \mathbf{Q}\mathbf{G}\left(\mathbf{G}^T\mathbf{Q}\mathbf{G}\right)^{-1}\mathbf{e}, \qquad [3.40]$$

$$\mathbf{P} = \mathbf{I} - \mathbf{Q}\mathbf{G}\left(\mathbf{G}^T\mathbf{Q}\mathbf{G}\right)^{-1}\mathbf{G}^T. \qquad [3.41]$$

For most problems, the simplest choice $\mathbf{Q} = \mathbf{I}$ is the most computationally efficient, and many studies reported in the literature have been performed with this choice (Rixen *et al.* 1999; Stefanica 2005). We apply the same treatment herein. Interestingly, the projection step involves the resolution of a coarse problem identified as $(\mathbf{G}^T\mathbf{Q}\mathbf{G})^{-1}$. This coarse problem is a dense system of the same size than the vector $\boldsymbol{\alpha}$ (which contains the amplitudes of the rigid-body motions). It links all the sub-domains together. It can be seen as a coarse correction that plays a particular role in the scalability of the algorithm.

Finally, full or partial reorthogonalization of the gradient at every iteration is commonly adopted to overcome the poor conditioning of the dual Schur complement (Gosselet and Rey 2006). In this work, a full reorthogonalization is performed as presented in Table 3.1.

Initialize	
$\boldsymbol{\lambda}^{(0)} = \mathbf{QG}(\mathbf{G}^T\mathbf{QG})^{-1}\mathbf{e}$	
$\mathbf{r}^{(0)} = \mathbf{d} - \mathbf{F}\boldsymbol{\lambda}^{(0)}$	
Iterate $k = 1, 2, \ldots$ until convergence	
$\mathbf{w}^{(k-1)} = \mathbf{P}^T\mathbf{r}^{(k-1)}$	Project
$\mathbf{z}^{(k-1)} = \mathbf{M}\mathbf{w}^{(k-1)}$	Precondition
$\mathbf{y}^{(k-1)} = \mathbf{P}\mathbf{z}^{(k-1)}$	Re-project
$\mathbf{p}^{(k)} = \mathbf{y}^{(k-1)} - \sum_{i=1}^{k-1} \beta^{(i)}\mathbf{p}^{(i)}$ $\quad (\mathbf{p}^{(1)} = \mathbf{y}^{(0)})$	Conjugate
with $\beta^{(i)} = \dfrac{\mathbf{y}^{(k-1)} \cdot \mathbf{F}\mathbf{p}^{(i)}}{\mathbf{p}^{(i)} \cdot \mathbf{F}\mathbf{p}^{(i)}}$	
$\alpha^{(k)} = \dfrac{\mathbf{y}^{(k-1)} \cdot \mathbf{w}^{(k-1)}}{\mathbf{p}^{(k)} \cdot \mathbf{F}\mathbf{p}^{(k)}}$	Minimize
$\boldsymbol{\lambda}^{(k)} = \boldsymbol{\lambda}^{(k-1)} + \alpha^{(k)}\mathbf{p}^{(k)}$	Update
$\mathbf{r}^{(k)} = \mathbf{r}^{(k-1)} - \alpha^{(k)}\mathbf{F}\mathbf{p}^{(k)}$	Residual

Table 3.1. *The preconditioned conjugate projected gradient algorithm with reorthogonalization. Operators $\mathbf{P}$ and $\mathbf{M}$ denote the projector and the preconditioner, respectively*

3.4.2.2. *Parallel computing*

Let us now highlight the interest of such algorithms for parallel computing. To understand this point, we can observe what happens during the product between a given vector $\mathbf{y}$ and the dual Schur complement $\mathbf{F}$:

$$\mathbf{F}\,\mathbf{y} = \left[\sum_s \mathbf{C}_s^F \mathbf{K}_s^+ \mathbf{C}_s^{F^T}\right]\mathbf{y} = \sum_s \left(\mathbf{C}_s^F \mathbf{K}_s^+ \mathbf{C}_s^{F^T}\mathbf{y}\right). \qquad [3.42]$$

It results in a sum of local contributions that can be computed independently and thus in parallel. Furthermore, each local contribution involves the local pseudo-inverse $\mathbf{K}_s^+$ and the coupling matrix $\mathbf{C}_s^F$ which remain unchanged during the iterative resolution, and are therefore built once and for all in a pre-processing step. More precisely, each local contribution is computed as follows:

Sub – domain

	step 1		step 2		step 3	
$\mathbf{y} \longrightarrow$	$\mathbf{v}_s^1 = \mathbf{C}_s^{F^T}\mathbf{y}$	$\longrightarrow$	$\mathbf{v}_s^2 = \mathbf{K}_s^+\mathbf{v}_s^1$	$\longrightarrow$	$\mathbf{v}_s^3 = \mathbf{C}_s^F\mathbf{v}_s^2$	$\longrightarrow \mathbf{v}_s^3$

As a result, a sub-domain can be seen as a black box: taking as an input a vector $\mathbf{y}$, it returns the local contribution $\mathbf{v}_s^3$. Summing all these local contributions gives the desired product $\mathbf{F}\mathbf{y}$. The most time-consuming step is the product by the pseudo-inverse (i.e. step 2). As will be discussed in the next section, the pseudo-inverse is generally given as a factorization (as a classical inverse). As a result, this step can be seen as the resolution of a linear system where the matrix is already factorized. Finally, let us note that a sub-domain does not need to know all the Lagrange multiplier DOF, but only those coming from the Lagrange multipliers that are actually acting on this sub-domain. Thus, a vector $\mathbf{y}$ is not communicated completely to each sub-domain, but only the active part $\mathbf{y}_s$ is required, which is extracted using trace operators. For example, taking the sub-domain Ω_1 from the case shown in Figure 3.6, we understand that only the DOF coming from the Lagrange multiplier $\boldsymbol{\lambda}_{12}$ is required. It also means that, in practice, the coupling matrix $\mathbf{C}_s^F$ does not have a number of rows equal to the total number of Lagrange multiplier DOF, but only equal to those related to the considered sub-domain.

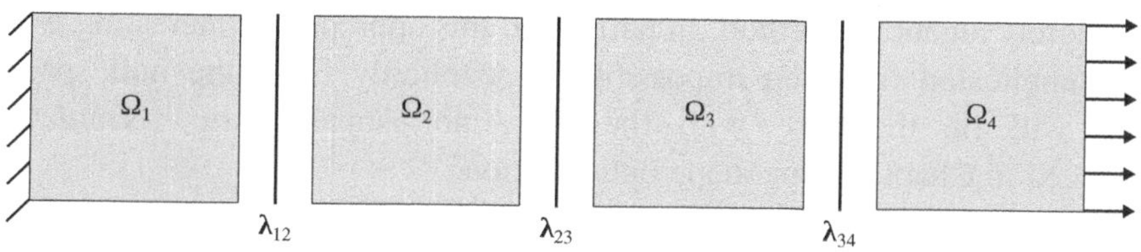

Figure 3.6. *Parallel computing: the communication between the sub-domains is performed through the Lagrange multipliers*

Unlike the dual Schur complement, there is, however, a step that occurs during the projection that can apparently not be done in parallel (see the algorithm described in Table 3.1). Indeed, in the expression of the projector [3.41], there is the coarse problem as mentioned earlier. However, as it takes the rigid-body DOF as unknowns, it is therefore of low dimension and can be easily factorized before entering into the gradient loop. For more details on parallel computing, interested readers can find a large amount of information in the literature since DDMs have been intensely developed over the past decades (see, for example, Toselli and Widlund (2006), Kozubek *et al.* (2013) and Dolean *et al.* (2015)).

3.4.3. *Null space and pseudo-inverse*

Up until now, we have not mentioned how the null spaces $\mathbf{R}_s$ and the pseudo-inverses $\mathbf{K}_s^+$ that come into play in the DD solver are formulated. To this end, there are different strategies, including the following two extremes: (i) a purely analytical or (ii) purely algebraic method. In approach (i), which is also called the Total FETI (Dostál *et al.* 2006), the starting point is to build the expression of the null space in the case of a floating sub-domain without any Dirichlet boundary conditions. This method partly aims to simplify the implementation of the original FETI method since the same treatment for each sub-domain is performed. Then, it is suggested to construct the pseudo-inverses by a regularization of the semi-definite stiffness matrices, which can be performed again through the use of the rigid-body modes (see, for example, Savenkov *et al.* (2008) and Kozubek *et al.* (2013) for more details). In approach (ii), the pseudo-inverses and the null spaces are constructed during the factorization of the stiffness matrices. In order to be general and independent from the starting formulation, we resort in this work to the purely algebraic method. In particular, this approach is interesting when it is complicated (or even impossible) to explicitly build the null spaces, which will be the case with the new, non-isoparametric formulation introduced in Chapter 4 for shape optimization.

For clarity, we voluntarily skip here the sub-domain identification through the subscript s, but the reader should view the following operators as local quantities. The purely algebraic method implemented in this work has been presented in Farhat *et al.* (1998), and was already used in Farhat and Roux (1991). The construction of the pseudo-inverse and null space is based on the partition of the matrix $\mathbf{K} \in \mathbb{R}^{n \times n}$ as:

$$\mathbf{K} = \begin{bmatrix} \mathbf{K}_{11} & \mathbf{K}_{12} \\ \mathbf{K}_{12}^T & \mathbf{K}_{22} \end{bmatrix}, \qquad [3.43]$$

where $\mathbf{K}_{11} \in \mathbb{R}^{r \times r}$ is a submatrix with the full rank r equal to the rank of the initial matrix (i.e. $r = \text{rk}(\mathbf{K}) = \text{rk}(\mathbf{K}_{11})$). With such a partition in hand, the following relation holds:

$$\mathbf{K}_{22} - \mathbf{K}_{12}^T \mathbf{K}_{11}^{-1} \mathbf{K}_{12} = \mathbf{0}. \qquad [3.44]$$

We can then define the pseudo-inverse and the null space as:

$$\mathbf{K}^{+} = \begin{bmatrix} \mathbf{K}_{11}^{-1} & \mathbf{0} \\ \mathbf{0} & \mathbf{0} \end{bmatrix}, \qquad \mathbf{R} = \begin{bmatrix} -\mathbf{K}_{11}^{-1}\mathbf{K}_{12} \\ \mathbf{I} \end{bmatrix}, \qquad [3.45]$$

where $\mathbf{I}$ is now the identity matrix of size $m = n - r$.

In practice, partition [3.43] is not known in advance, but it is obtained in an unordered form during a Gaussian-based factorization (as, for example, an LU (lower-upper) decomposition). At each step of the factorization, we can verify whether or not a null pivot is encountered. If verified, the corresponding equation is redundant and removed from the system. The reduced column can be recycled to recover the null space later.

3.4.4. *Preconditioning*

Last but not least, the ingredient involved in our algorithm is the preconditioner (already introduced in Table 3.1 through the operator $\mathbf{M}$). A good preconditioner is crucial for the computational performance of the iterative resolution. Its goal is to reduce the conditioning number of the operator $\mathbf{F}$ since the performance of an iterative Krylov solver is known to be highly dependent on the conditioning of the system to be solved. It can thus be seen as an approximation of the inverse of the system. The better the approximation, the more efficient the preconditioning step during the iterative resolution. However, it has to be cheap in order to be built and evaluated within the algorithm.

3.4.4.1. *Dirichlet preconditioner*

Several preconditioners of the dual Schur complement have been studied in the literature. In the context of FEM matching grids, the Dirichlet preconditioner is said to be optimal regarding the asymptotic bound and thus offers an excellent numerical scalability (Rixen 2002). The Dirichlet preconditioner is obtained under the approximation that the inverse of the sum of the local contributions is the sum of the inverses of each of these local contributions. More precisely, it is obtained by exactly inverting the local dual Schur complements and summing each local contribution. As justified below, in the case of a Mortar interface, we can rewrite this Dirichlet preconditioner as:

$$\tilde{\mathbf{F}}_I^{-1} = \sum_{s=1}^{n_s} \left(\mathbf{C}_s^F \, \mathbf{C}_s^{F^T}\right)^{-1} \mathbf{C}_s^F \begin{bmatrix} \mathbf{0} & \mathbf{0} \\ \mathbf{0} & \mathbf{S}_{s_{bb}} \end{bmatrix} \mathbf{C}_s^{F^T} \left(\mathbf{C}_s^F \, \mathbf{C}_s^{F^T}\right)^{-1}, \quad [3.46]$$

where the matrix $\mathbf{S}_{s_{bb}}$ denotes the local primal Schur complement, which is given by:

$$\mathbf{S}_{s_{bb}} = \mathbf{K}_{s_{bb}} - \mathbf{K}_{s_{bi}} \mathbf{K}_{s_{ii}}^{-1} \mathbf{K}_{s_{ib}}. \quad [3.47]$$

The subscript b corresponds to the boundary DOF on the interface and the subscript i is associated with the internal DOF.

PROOF.– *The expression of the Dirichlet preconditioner in the case of non-conforming interfaces may not be obvious. Let us thus prove now that expression* [3.46] *corresponds to the sum of the inverses of the local dual Schur complements. To this purpose, we can rewrite the local dual Schur complement* $\mathbf{F}_s = \mathbf{C}_s^F \mathbf{K}_s^{-1} \mathbf{C}_s^{F^T}$ *(see equation* [3.33], *where we consider the case of* $\mathbf{K}_s^+ = \mathbf{K}_s^{-1}$ *for simplicity) by using the splitting of the boundary DOF and the interface DOF. It can be written as:*

$$\mathbf{F}_s = \begin{bmatrix} 0 & \mathbf{C}_{s_b}^F \end{bmatrix} \begin{bmatrix} \mathbf{K}_{s_{ii}} & \mathbf{K}_{s_{ib}} \\ \mathbf{K}_{s_{bi}} & \mathbf{K}_{s_{bb}} \end{bmatrix}^{-1} \begin{bmatrix} 0 \\ \mathbf{C}_{s_b}^{F^T} \end{bmatrix} \quad [3.48]$$

Next, we introduce the following quantities $\mathbf{V}_{s_i}$ *and* $\mathbf{V}_{s_b}$ *such that:*

$$\begin{bmatrix} \mathbf{V}_{s_i} \\ \mathbf{V}_{s_b} \end{bmatrix} = \begin{bmatrix} \mathbf{K}_{s_{ii}} & \mathbf{K}_{s_{ib}} \\ \mathbf{K}_{s_{bi}} & \mathbf{K}_{s_{bb}} \end{bmatrix}^{-1} \begin{bmatrix} 0 \\ \mathbf{C}_{s_b}^{F^T} \end{bmatrix}. \quad [3.49]$$

Thus, we can rewrite equation [3.48] *by using the notation from equation* [3.49]*:*

$$\mathbf{F}_s = \begin{bmatrix} 0 & \mathbf{C}_{s_b}^F \end{bmatrix} \begin{bmatrix} \mathbf{V}_{s_i} \\ \mathbf{V}_{s_b} \end{bmatrix} = \mathbf{C}_{s_b}^F \mathbf{V}_{s_b} \Rightarrow \mathbf{V}_{s_b} = \mathbf{C}_{s_b}^{F^T} \left(\mathbf{C}_s^F \, \mathbf{C}_s^{F^T}\right)^{-1} \mathbf{F}_s. \quad [3.50]$$

We can multiply equation [3.49] *on the left by the stiffness matrix* $\mathbf{K}_s$. *It yields the following equations:*

$$\mathbf{K}_{s_{ii}}\mathbf{V}_{s_i} + \mathbf{K}_{s_{ib}}\mathbf{V}_{s_b} = 0, \tag{3.51}$$

$$\mathbf{K}_{s_{bi}}\mathbf{V}_{s_i} + \mathbf{K}_{s_{bb}}\mathbf{V}_{s_b} = \mathbf{C}_{s_b}^{F\,T}. \tag{3.52}$$

Now, we substitute equation [3.51] *into equation* [3.52]. *The local primal Schur complement, as previously defined in equation* [3.47], *is:*

$$\left[\mathbf{K}_{s_{bb}} - \mathbf{K}_{s_{bi}}\,\mathbf{K}_{s_{ii}}^{-1}\mathbf{K}_{s_{ib}}\right]\mathbf{V}_{s_b} = \mathbf{C}_{s_b}^{F\,T} \quad \Rightarrow \quad \mathbf{S}_{s_{bb}}\mathbf{V}_{s_b} = \mathbf{C}_{s_b}^{F\,T}. \tag{3.53}$$

Introducing the result from equation [3.50] *into equation* [3.53] *gives:*

$$\mathbf{S}_{s_{bb}}\mathbf{C}_{s_b}^{F\,T}\left(\mathbf{C}_s^F\,\mathbf{C}_s^{F\,T}\right)^{-1}\mathbf{F}_s = \mathbf{C}_{s_b}^{F\,T} \tag{3.54}$$

Finally, we multiply this last equation on the left by the term $\left(\mathbf{C}_s^F\,\mathbf{C}_s^{F\,T}\right)^{-1}\mathbf{C}_{s_b}^F$. *We obtain the final result:*

$$\left(\mathbf{C}_s^F\,\mathbf{C}_s^{F\,T}\right)^{-1}\mathbf{C}_{s_b}^F\mathbf{S}_{s_{bb}}\mathbf{C}_{s_b}^{F\,T}\left(\mathbf{C}_s^F\,\mathbf{C}_s^{F\,T}\right)^{-1}\mathbf{F}_s = \mathbf{I}_b \tag{3.55}$$

We find the expression presented in equation [3.46].

3.4.4.2. *Generalized preconditioner*

In the particular case of Mortar coupling, some improvements of the initial version of the Dirichlet preconditioner have been studied (Lacour 1997; Stefanica 2001; Rixen 2002; Stefanica 2005). For example, scaling factors can be added to take into account the possible gap between the stiffnesses on both sides of the interface (Rixen and Farhat 1999; Klawonn and Widlund 2001). These heterogeneities are particularly numerous in the case of different material properties, stiffened structures with the presence of T-junctions, different shell thicknesses, etc. (Amini *et al.* 2009). Thus, it is primordial to efficiently deal with these heterogeneities to analyze complex structures. To this end, it is proposed here to follow the extended Dirichlet preconditioner suggested by Rixen (2002) and mentioned by Stefanica (2005), which can be written as:

$$\tilde{\mathbf{F}}_A^{-1} = \sum_{s=1}^{n_s}\left(\mathbf{C}^F\mathbf{A}\mathbf{C}^{F\,T}\right)^{-1}\mathbf{C}_s^F\mathbf{A}_s\begin{bmatrix}\mathbf{0} & \mathbf{0}\\ \mathbf{0} & \mathbf{S}_{s_{bb}}\end{bmatrix}\mathbf{A}_s\mathbf{C}_s^{F\,T}\left(\mathbf{C}^F\mathbf{A}\mathbf{C}^{F\,T}\right)^{-1}, \tag{3.56}$$

where $\mathbf{C}^F$ is the global coupling operator of equation [3.25] and the matrix $\mathbf{A}_s$ is chosen here as the superlumped scaling: $\mathbf{A}_s = \mathrm{diag}(\mathbf{K}_s)^{-1}$. The global matrix $\mathbf{A}$ is the diagonal block assembly of all these local scaling matrices, which leads to a diagonal matrix in the present case.

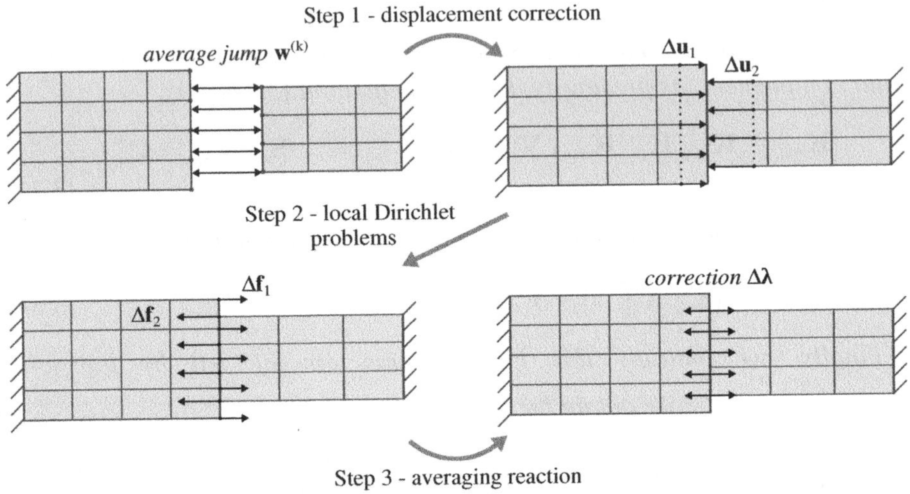

Figure 3.7. *Illustration of the preconditioning step (generalized preconditioner) during the iterative resolution of the dual domain decomposition solver (adapted from Gosselet and Rey (2006)). For a color version of this figure, see www.iste.co.uk/bouclier/IGA.zip*

The extended Dirichlet preconditioner has shown better performance in comparison with the original Dirichlet preconditioner in the case of Mortar interfaces (Lacour 1997; Stefanica 2001, 2005). Rixen (2002) shows that this preconditioner is mechanically consistent and mentions its mechanical meaning: once applied to the residual of the interface problem [3.28], it computes the correction for the Lagrange multipliers. Figure 3.7 highlights this point. More precisely, we can split the preconditioner into three steps to understand its mechanical meaning:

1) **Make the interface compatibility** [3.28] **satisfied.** Let $\mathbf{u}^{(k)}$ and $\mathbf{w}^{(k)}$ be, respectively, the displacement and the residual at iteration k (i.e. $\mathbf{C}^F\mathbf{u}^{(k)} = \mathbf{w}^{(k)}$). We are looking for the displacement correction $\Delta\mathbf{u}$ such that:

$$\mathbf{C}^F(\mathbf{u}^{(k)} - \Delta\mathbf{u}) = \mathbf{0}, \quad \text{which gives} \quad \mathbf{C}^F\Delta\mathbf{u} = \mathbf{w}^{(k)}. \qquad [3.57]$$

Thus, we can build the correction, in a least-square sense, as:

$$\Delta\mathbf{u} = \mathbf{A}\mathbf{C}^{F^T}\left(\mathbf{C}^F\mathbf{A}\mathbf{C}^{F^T}\right)^{-1}\mathbf{w}^{(k)}, \qquad [3.58]$$

where the matrix $\mathbf{A}$ is invertible.

2) **Compute the interface efforts corresponding to the jump correction.** It is performed at the sub-domain level through the local primal Schur complements $\mathbf{S}_s$:

$$\Delta\mathbf{f}_s = \mathbf{S}_s\Delta\mathbf{u}_s. \qquad [3.59]$$

3) **Make the Lagrange multipliers represent these interface efforts.** We are looking for the correction $\Delta\boldsymbol{\lambda}$ such that:

$$\mathbf{C}^{F^T}\Delta\boldsymbol{\lambda} = \Delta\mathbf{f}. \qquad [3.60]$$

As for step 1, this equation can be satisfied in a least-square sense:

$$\Delta\boldsymbol{\lambda} = \left(\mathbf{C}^F\mathbf{A}\mathbf{C}^{F^T}\right)^{-1}\mathbf{C}^F\mathbf{A}\,\Delta\mathbf{f}. \qquad [3.61]$$

Stefanica (2005) discusses the parallelization properties of this generalized preconditioner. The critical point concerns the two global multiplications by $\left(\mathbf{C}^F\,\mathbf{A}\,\mathbf{C}^{FT}\right)^{-1}$ which cannot be performed in parallel. We can build local matrices $\tilde{\mathbf{C}}^F = \left(\mathbf{C}^F\,\mathbf{A}\,\mathbf{C}^{FT}\right)^{-1}\mathbf{C}^F_s$ once and for all at the beginning of the resolution. In order to preserve the sparsity, biorthogonal mortars could be helpful in this case (Stefanica (2005); Wunderlich *et al.* (2019)).

3.5. Numerical investigation of the developed algorithms

We eventually run several examples to highlight the good behavior of the constructed DD algorithms. We note that only a sequential implementation has been performed so far. The goal here is to make sure that the algorithm can be envisaged for HPC. The true parallel implementation does not raise any difficulty from a theoretical point of view; only some coding precautions should be taken to fully benefit from the efficiency of the method (see Toselli and Widlund (2006), Kozubek *et al.* (2013) and Dolean *et al.* (2015) for technical aspects regarding parallel computing).

First, we place ourselves in the context of standard 2D elasticity; then, we investigate the bending of a heterogeneous plate and the famous Scordelis–Lo roof benchmark is revisited in a DD context; finally, a quite complex stiffened panel is computed. At this stage, let us mention that a more complex test case representative of an aircraft wing substructure will be computed and optimized in Chapter 4 (see section 4.6.3). The convergence criteria are defined as a relative tolerance on the projected residual (namely $\mathbf{w}^{(k)}$ in Table 3.1 set as $||\mathbf{w}^{(k)}|| < 10^{-8}||\mathbf{w}^{(0)}||$). For every example, we depict the number of iterations versus the level of refinement of the mesh given a fixed decomposition. The results are indicated in tables, where the headers n_{tot} and n_λ denote the number of total and interface DOF, respectively (i.e. by referring to [3.25], n_λ is the size of $\boldsymbol{\lambda}$ and n_{tot} is the total number of unknowns). The columns with the headers $\mathbf{I}$, $\tilde{\mathbf{F}}_I^{-1}$ and $\tilde{\mathbf{F}}_A^{-1}$ give the iteration number without any preconditioner, with the Dirichlet preconditioner [3.46] and with the generalized preconditioner [3.56], respectively. For the sake of clarity, we omit to report the convergence study for the classical Dirichlet preconditioner when it leads to identical results for the generalized preconditioner. This is especially the case for homogeneous domains where both presented preconditioners [3.46] and [3.56] improve the convergence rate in approximately the same manner. The goal of these tables is to provide insights into the numerical scalability of the corresponding DD algorithm. Furthermore, to show the accuracy of the numerical procedure, we study the convergence of the final iterative solution with respect to mesh refinement for the two homogeneous examples, i.e. the 2D cantilever beam and the Scordelis–Lo roof. In order to do so, we consider the relative energy error as:

$$\text{Err}^h = \frac{|E_{\text{ref}} - E^h_{\text{coupling}}|}{E_{\text{ref}}}, \qquad [3.62]$$

where the reference energy E_{ref} is computed on a very fine single-patch discretization.

3.5.1. *Standard solid elasticity*

Two plane stress problems are considered to assess the performance in the context of solid elasticity. As stated previously, we start here with the linear system of form [2.1] as a Mortar coupling formulation. Only the displacement is thus enforced thanks to an additional Lagrange multiplier field. For the discretization of the Lagrange multiplier field, we perform similarly as with our Kirchhoff–Love shell coupling scheme of section 3.3.2, i.e. we take a vector-valued spline function with degree $p-1$ and make sure to take a mesh at least as coarse as the coarsest of the sub-domains to be coupled. Then, we apply the strategy described in section 3.4 to build the DD algorithm.

3.5.1.1. *Homogeneous 2D cantilever beam*

To begin with, we investigate the bending of a 2D cantilever beam with a homogeneous material, as described in Figure 3.8. Three different configurations are studied, and these decompositions are denoted as:

– `1A`: two non-matching patches with a similar geometry;

– `1B`: four non-matching patches with a similar geometry;

– `1C`: four non-matching patches with different geometries.

For every configuration, the patches are discretized with cubic splines.

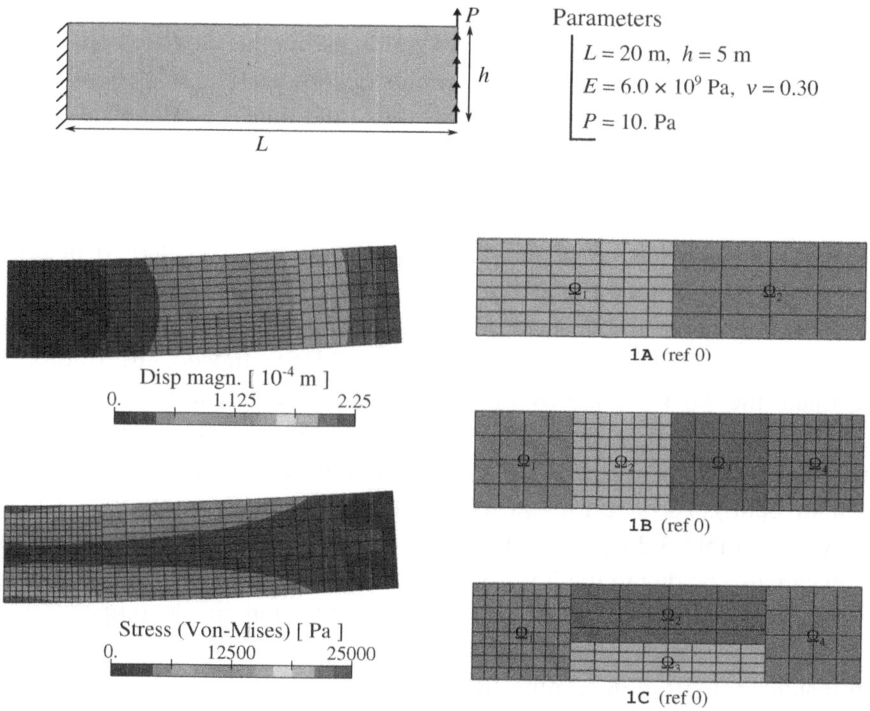

Figure 3.8. *Description and results of the homogeneous 2D cantilever beam problem. Three decompositions 1A, 1B and 1C are studied. The level of refinement (ref i) is given according to Table 3.2. For a color version of this figure, see www.iste.co.uk/bouclier/IGA.zip*

	1A					1B					1C				
ref	n_{tot}	n_λ	**I**	$\tilde{\mathbf{F}}_A^{-1}$	Err^h	n_{tot}	n_λ	**I**	$\tilde{\mathbf{F}}_A^{-1}$	Err^h	n_{tot}	n_λ	**I**	$\tilde{\mathbf{F}}_A^{-1}$	Err^h
0	330	12	4	4	4.86e-4	702	36	13	8	5.11e-4	630	60	48	13	1.94e-4
1	946	20	9	7	1.64e-4	1966	60	19	9	1.84e-4	1686	100	85	14	6.63e-5
2	3138	36	21	9	5.39e-5	6414	108	36	10	6.37e-5	5334	180	149	15	2.07e-5
3	11 362	68	45	9	1.61e-5	22 990	204	64	10	2.01e-5	18 774	340	251	15	4.60e-6
4	43 170	132	71	9	2.92e-6	86 862	396	95	10	4.44e-6	70 230	660	347	16	1.05e-6

Table 3.2. *Performance study for the homogeneous cantilever beam problem described in Figure 3.8. The reference energy was taken as* $E_{\text{ref}} = 5.568207 \cdot 10^{-2}$ *to compute the relative energy error of the coupled solution (see equation* [3.62]*)*

As can be seen in Figure 3.8, the displacement and stress fields are smooth and clearly express the bending of the beam. Then, Table 3.2 summarizes the convergence results for the DD algorithm for this first test case. We can see that the algorithm requires very few iterations to reach convergence when the preconditioner [3.56] is used. The gain is all the more significant in comparison with the unpreconditioned version as fine meshes are considered. This is especially true for the more complex decomposition 1C. These results demonstrate the good convergence properties (and thus the efficiency) of our DD solver.

NOTE.– It may be noted that since we consider a fixed decomposition for studying the performance of our DD solver, it is not expected to obtain a constant number of iterations in the different tables of this section, even if our algorithm is scalable in the sense of Farhat *et al.* (1994), Rixen *et al.* (1999), Jolivet *et al.* (2014) and Klawonn *et al.* (2015). Indeed, to rigorously investigate the weak scalability of the algorithm, we should have refined the mesh along with the decomposition to satisfy a constant sub-domain-to-element size ratio (the famous H/h ratio) (Toselli and Widlund 2006). As a result, the constant number of iterations that can be observed in Table 3.2 does confirm the good performance of our algorithm, which can also be due to the low complexity of the proposed example. Finally, we insist again that for general applications, the domain decomposition in IGA has a geometrical meaning in the sense that it is provided by the multipatch configuration coming from CAD. This partly explains why we did not modify the domain decomposition in this study (as in the original DD work in IGA (Kleiss *et al.* 2012)). This statement makes even more sense in the case of stiffened structures, as studied in section 3.5.4 for example.

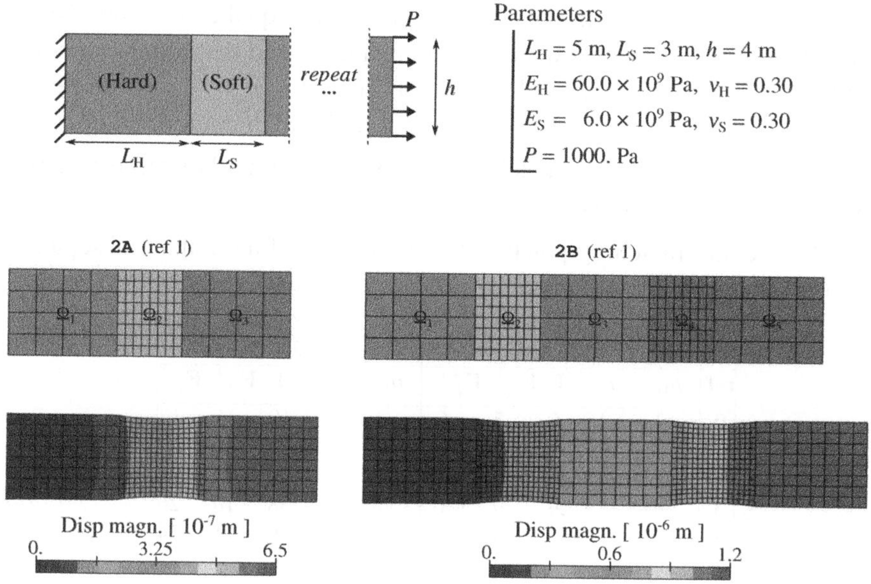

Figure 3.9. *Description and results of the heterogeneous 2D cantilever beam problem. The beam is made up of a sequence of hard and soft regions. Configuration 2A has two hard regions separated by one soft region. Configuration 2B has three hard regions separated by two soft regions. The level of refinement (ref i) is given according to Table 3.3. For a color version of this figure, see www.iste.co.uk/bouclier/IGA.zip*

Finally, we can also observe the results regarding the accuracy of the approach in the columns with the header Err^h in Table 3.2. The relative energy error [3.62] decreases with the mesh refinement. It indicates that the solutions obtained with the presented solver for the different multipatch configurations converge to the single-patch solution (reference solution). We observe the same convergence rate for each configuration. More interestingly, it can be noted that this convergence rate is identical to that of the single-patch discretization.

3.5.1.2. *Heterogeneous 2D cantilever beam*

The second test case consists of a heterogeneous 2D beam. The description of the problem is given in Figure 3.9. The role of the preconditioner is known to be crucial when the sub-domains have large differences of material properties (Rixen and Farhat 1999; Gosselet *et al.* 2015; Bovet *et al.* 2017). The beam is made up of a sequence of hard and soft regions. One side of the beam is fixed and the other side is subjected to a traction load. Each hard region is defined by a 5 m by 4 m rectangle and has a Young modulus equal to $E = 60 \cdot 10^9$ Pa. Each soft region is defined by a 3 m by 4 m rectangle and has a Young modulus equal to $E = 6 \cdot 10^9$. The same Poisson ratio equal to $\nu = 0.30$ is chosen for both regions. We consider two cases of cubic multipatch configurations:

– 2A: three non-matching patches: two made up of hard material separated by one made up of soft material.

– 2B: five non-matching patches; three made up of hard material separated by two made up of soft material.

	2A					2B				
ref	n_{tot}	n_λ	**I**	$\tilde{\mathbf{F}}_A^{-1}$	$\tilde{\mathbf{F}}_I^{-1}$	n_{tot}	n_λ	**I**	$\tilde{\mathbf{F}}_A^{-1}$	$\tilde{\mathbf{F}}_I^{-1}$
0	204	16	6	5	6	368	32	10	6	12
1	448	24	9	7	10	812	48	14	8	16
2	1224	40	15	9	15	2228	80	22	10	21
3	3928	72	29	9	21	7172	144	38	10	26
4	13 944	136	55	10	27	25 508	272	69	11	34

Table 3.3. *Performance study for the heterogeneous beam problem described in Figure 3.9*

Table 3.3 gives the iterative convergence results for the present case. As for the previous example of the homogeneous cantilever beam, the algorithm requires very few iterations to reach convergence when the preconditioner $\tilde{\mathbf{F}}_A^{-1}$ is used, and the gain compared to the unpreconditioned version (columns with header **I**) becomes substantial for fine meshes. Therefore, the heterogeneity is correctly tackled by the generalized preconditioner thanks to the superlumped scaling. We also give in Table 3.3 the convergence when the classical Dirichlet preconditioner $\tilde{\mathbf{F}}_I^{-1}$ is used. We can see that the benefit is not as good as we can expect from a preconditioner. In fact, for the low levels of refinement, there are no iteration savings when using the classical Dirichlet preconditioner in comparison with the unpreconditioned cases. For finer discretizations, it improves the convergence slowly. As an additional illustration, Figure 3.10 presents the convergence histories in the case of the configuration 2B with the highest level of refinement. For this particular case, the solver convergence is twice as fast with the preconditioner $\tilde{\mathbf{F}}_I^{-1}$, while the generalized Dirichlet preconditioner drastically improves the convergence. This exhibits again that the scaling during the preconditioning is essential to correctly balance the effort between heterogeneous sub-domains.

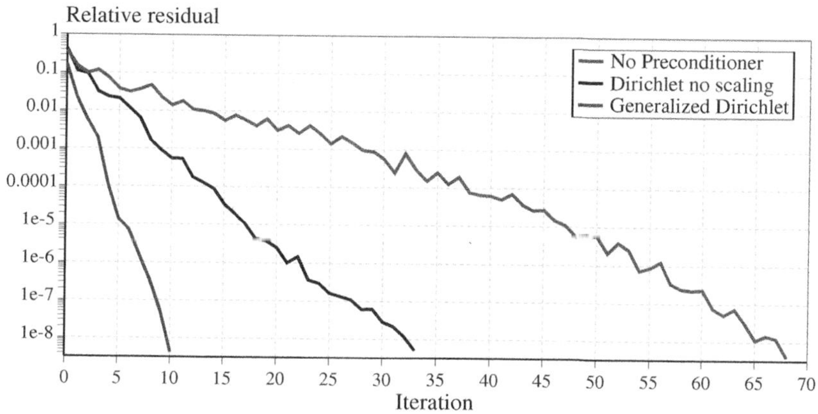

Figure 3.10. *Convergence of the dual DD solver when applied to the heterogeneous beam problem with five sub-domains (see configuration 2B in Figure 3.9) and a refinement level of 4 according to Table 3.3. The scaling step of the generalized Dirichlet preconditioner drastically improves the convergence rate. For a color version of this figure, see www.iste.co.uk/bouclier/IGA.zip*

3.5.2. *Heterogeneous plate bending*

We now switch to the computation of multipatch Kirchhoff–Love shell models. We begin here with the discrete Mortar scheme of equation [3.23] and then apply the strategy provided in section 3.4 to obtain the DD solver. The first example in this context consists of the bending of a heterogeneous plate, as described in Figure 3.11. The plate is fully fixed at one side (i.e. both in terms of displacements and rotations) while a uniform pressure is applied over the whole structure. We decompose the plate into two patches with different thicknesses such that the ratio is $t_1/t_2 = 0.1$. As a result, we address here a heterogeneous plate since its behavior (in particular its bending) continuously depends on its thickness. Cubic B-spline basis functions are used and the discretization of the patch with the fixed edge is twice finer than the second one. The geometric and material parameters for the plate are given in Figure 3.11.

The results in terms of bending moments are plotted in Figure 3.11 (bottom). Note that the Poisson ratio is not taken as 0 here. As a result, the x-component of the bending moment does not vanish, and it is discontinuous across Γ given the thickness gap between the two patches. Conversely, the y-component of the bending moment is continuous over the whole structure.

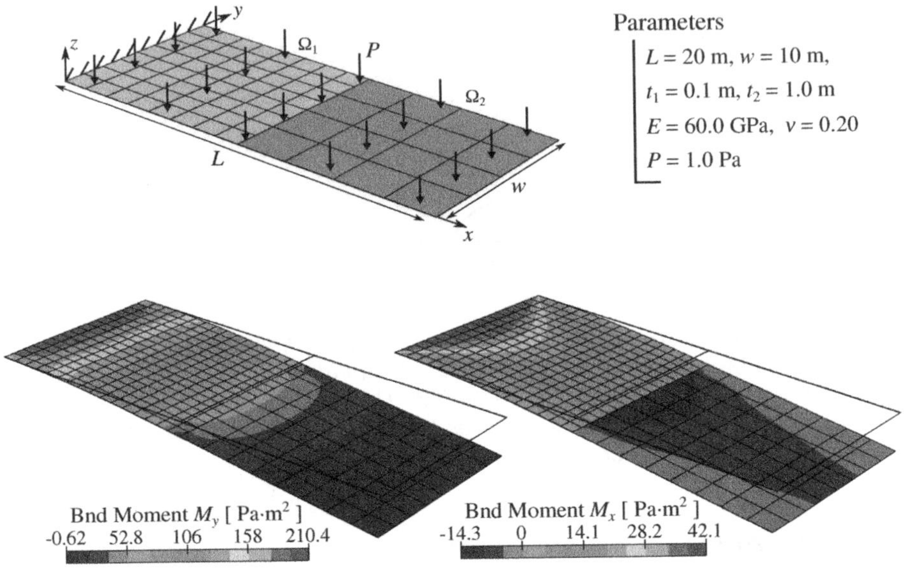

Figure 3.11. *Description and results of the bending plate problem. The plate is decomposed into two non-matching patches. We study a heterogeneous case: the two coupled shells have different thicknesses. For a color version of this figure, see www.iste.co.uk/bouclier/IGA.zip*

ref	n_{tot}	n_λ	**I**	$\tilde{\mathbf{F}}_A^{-1}$	$\tilde{\mathbf{F}}_I^{-1}$
0	222	28	7	4	7
1	510	44	14	4	11
2	1446	76	28	6	18
3	4758	140	59	11	28
4	17 142	268	117	16	33

Table 3.4. *Performance study for the bending plate problem described in Figure 3.11. Note: the mesh corresponding to the refinement level 2 is the one used to depict the bending moments in Figure 3.11 (top)*

In addition, Table 3.4 presents the performance study of the algorithm for this problem. As for the previous heterogeneous test case, we can see that the algorithm requires few iterations to reach convergence with a preconditioner and, more specifically, when the generalized preconditioner [3.56] is used. This also accounts for the correct tackling of the heterogeneity with $\tilde{\mathbf{F}}_A^{-1}$ in the case of shells. Furthermore, it can be observed that the increase in iterations between two mesh refinement steps is drastically reduced thanks to the help of a preconditioner. Consequently, for the finest refinement levels, the gain is significant in comparison with the unpreconditioned version.

NOTE.– The interested reader may also observe that the iterative convergence, even with $\tilde{\mathbf{F}}_A^{-1}$, is less impressive than in the 2D elasticity context (see Table 3.2). This may be due to the higher complexity of the numerical model, as well as the higher-order nature of the governing equations to be solved. In fact, the one-level FETI equipped with the Dirichlet preconditioner is known to be numerically scalable for second-order problems. However, it does not hold true for fourth-order problems, as shown, for example, by Farhat *et al.* (1998). This is why extensions of the classical FETI method, for example the two-level FETI and later on the FETI-DP, have been introduced to optimally deal with plates and shells (Farhat *et al.* 1998, 2001). Nevertheless, for this test case, the developed one-level FETI-based DD algorithm performs well and leads to entirely acceptable results in our opinion.

3.5.3. *Scordelis–Lo roof*

The Scordelis–Lo roof is part of the shell obstacle course that is widely used to study the performance of shell formulations (Belytschko *et al.* 1985). It consists of a portion of cylinders subjected to a vertical gravity load and fixed at its two end sections using rigid diaphragms. The problem parameters and some results for this test case can be found in the abundant shell literature (see, for example, Belytschko *et al.* (1985), Kiendl *et al.* (2009) and Bouclier *et al.* (2013a)). For our DD study, we decompose the Scordelis–Lo roof into two non-matching configurations, as shown in Figure 3.12 (left):

– `3A`: four quasi-identical cubic patches which intersect at a cross point;

– `3B`: nine patches with different levels of mesh refinement and polynomial degree. The coarsest are quartic patches, while the finest are cubic patches.

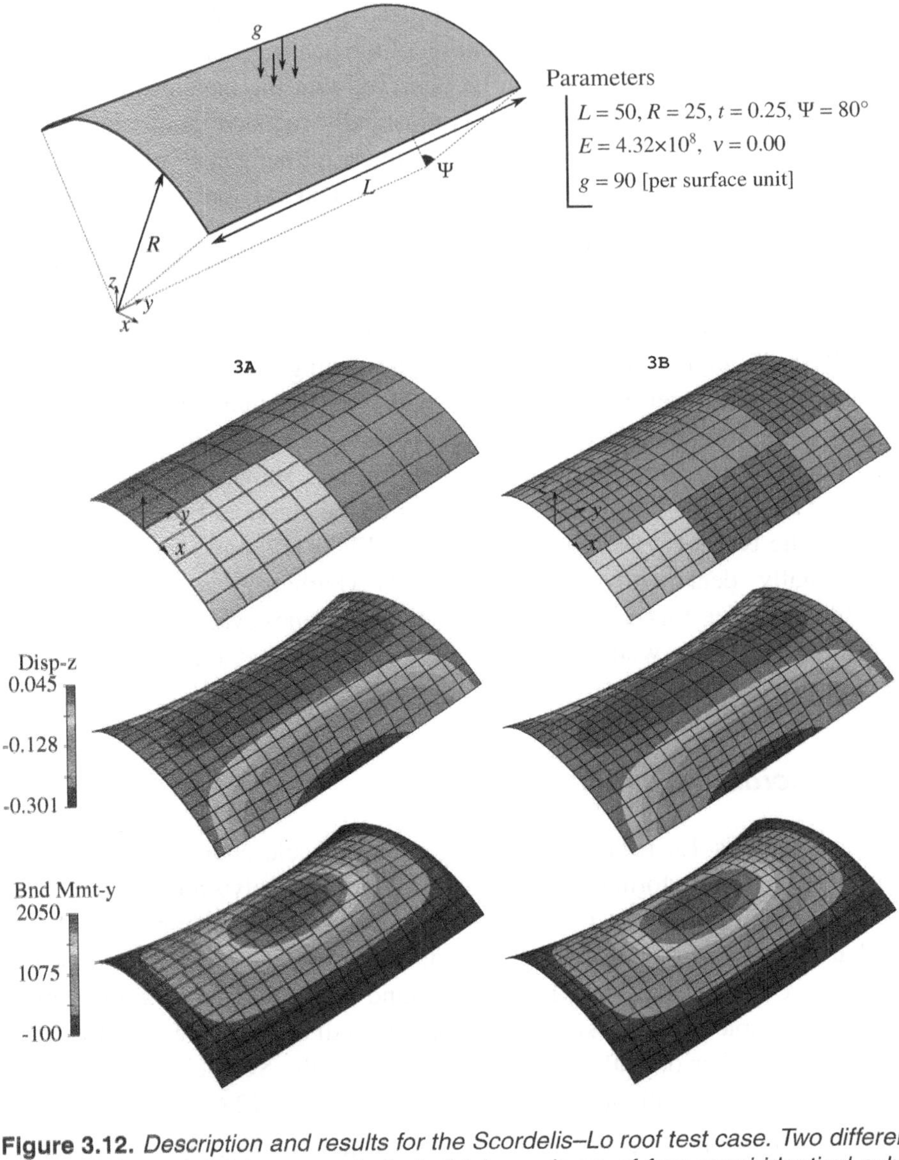

Figure 3.12. *Description and results for the Scordelis–Lo roof test case. Two different non-matching subdivisions are studied: 3A is made up of four quasi-identical cubic patches (two patches have one more element per direction) and 3B is made up of nine patches with different levels of mesh refinement and polynomial degree. Note: the input data and the results are given in a consistent unit system, according to Belytschko* et al. *(1985). For a color version of this figure, see www.iste.co.uk/bouclier/IGA.zip*

	3A					3B			
ref	n_{tot}	n_λ	**I**	$\tilde{\mathbf{F}}_A^{-1}$	Err^h	n_{tot}	n_λ	**I**	$\tilde{\mathbf{F}}_A^{-1}$
0	708	92	70	34	3.55e-3	1513	276	167	44
1	1652	156	117	38	6.44e-5	3017	468	317	52
2	4692	284	188	48	1.92e-6	7537	842	558	64
3	15 380	540	276	64	8.54e-8	22 625	1620	949	87
4	55 188	1052	405	86	3.52e-9	76 993	3156	1783	121

Table 3.5. *Performance study for the Scordelis–Lo roof shell problem described in Figure 3.12. Note: the refinement level 1 corresponds to the meshes in Figure 3.12 where the analysis results are presented*

Figure 3.12 shows the vertical displacement field and the distribution of a bending moment for each configuration, which are as smooth as expected. The results for the iterative convergence of the strategy are given in Table 3.5. It can be seen that the benefit of the preconditioning step increases from the first configuration 3A to the second configuration 3B. For the first configuration 3A, the iteration count is reduced by a factor of 4 for the refinement levels above 1. For the more complex decomposition 3B, the iteration count is drastically reduced when using the generalized Dirichlet preconditioner. Looking at the refinement levels 3 and 4 in Table 3.5 (and the columns referred to configuration 3B), we observe that the number of iterations is reduced by a factor of 10 and 15, respectively. Thus, the benefit provided by the preconditioner is all the more remarkable in the case of complex DD. Overall, the growth of the iteration count with the refinement level is very slow for each configuration, which leads to entirely satisfactory results.

In addition, the results in terms of energy errors [3.62] for decomposition 3A are provided in Table 3.5. It enables us to show that the coupling strategy presented in section 3.3 is valid. The relative energy error is low and decreases with the mesh refinement. In particular, this convergence rate is similar to the one we observe for the cubic monopatch discretization. Finally, note that we do not perform the convergence study for configuration 3B since it contains patches of different degrees (cubic and quartic).

3.5.4. *Stiffened panel*

Finally, the problem of the stiffened panel is presented in Figure 3.13 (top). It consists of the assembly of a square plate with subparts called the stiffeners. We design three stiffeners: two with a parabolic shape and the third one with a straight shape. The straight stiffener intersects its two curved counterparts.

As a result, five interfaces are defined, including two non-conforming ones (related to the curved stiffeners). Every patch is discretized using cubic Kirchhoff–Love shells. The material and geometric parameters of the study are shown in Figure 3.13, along with the results in terms of displacements and stress resultants. The edges of the panel are fixed (no displacement).

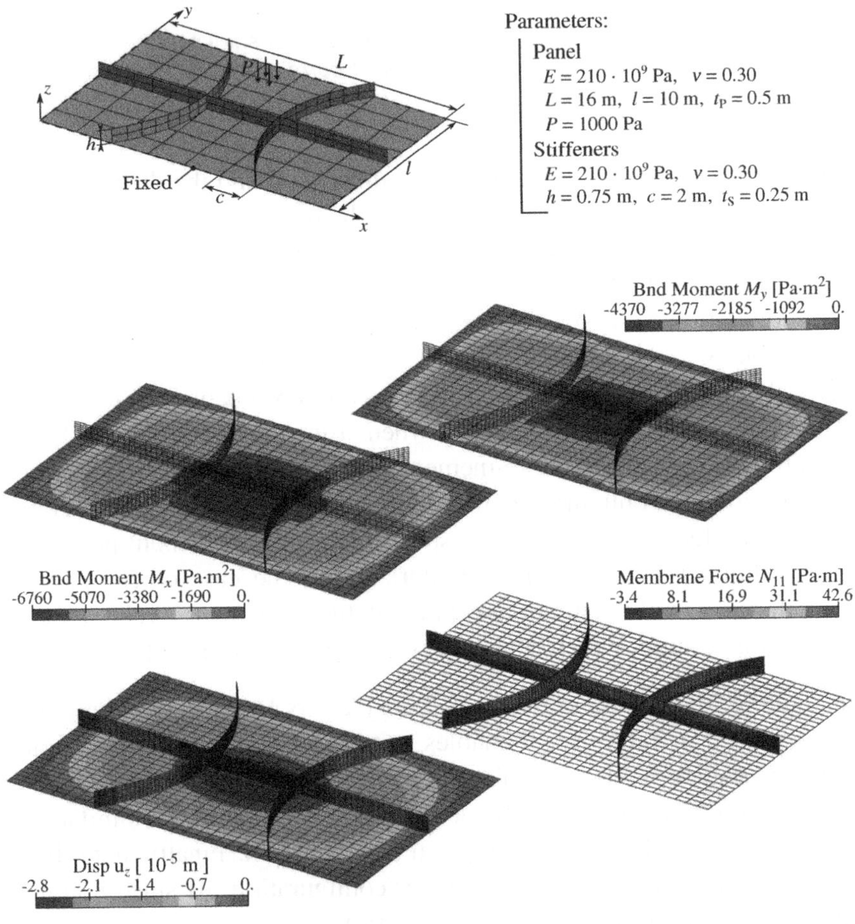

Figure 3.13. *Description and results for the stiffened panel problem. It consists of the assembly of a rectangular plate with three stiffeners (two with a parabolic shape crossed by a straight shape). It leads to four patches and five coupling interfaces (three stiffener/panel and two stiffener/stiffener interfaces). In particular, the two curved stiffeners lead to two non-conforming coupling interfaces. The main stress resultants for the panel are the bending moments, whereas the stiffeners are mainly subjected to membrane forces, which highlights the heterogeneity of the test case. For a color version of this figure, see www.iste.co.uk/bouclier/IGA.zip*

ref	n_{tot}	n_λ	**I**	$\tilde{\mathbf{F}}_A^{-1}$	$\tilde{\mathbf{F}}_I^{-1}$
0	1227	147	109	50	81
1	3313	259	193	68	94
2	10 365	483	341	85	110
3	35 989	931	613	103	137
4	133 317	1827	1106	125	189

Table 3.6. *Performance study for the stiffened panel problem described in Figure 3.13. Note: refinement level 0 corresponds to the situation of Figure 3.13 (top)*

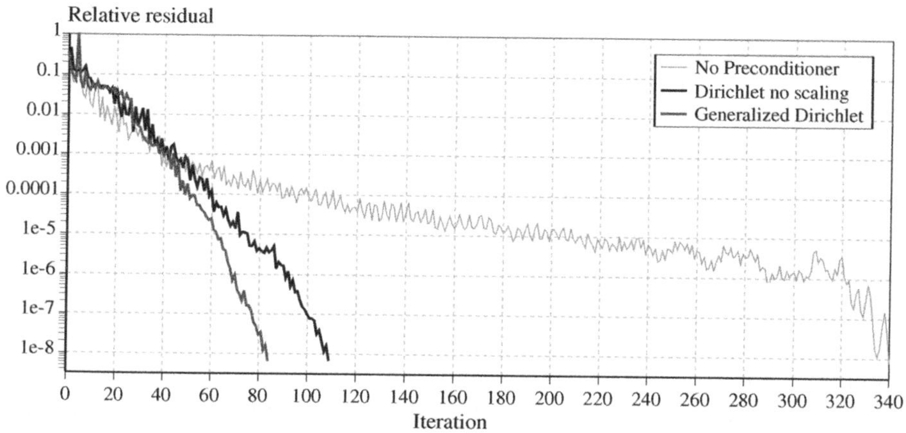

Figure 3.14. *Convergence plot of the dual DD solver when applied to the stiffened panel (refinement level 2 according to Table 3.6). Without the preconditioning step, the solver faces difficulties converging. The scaling within the generalized preconditioner accelerates the convergence in comparison with the Dirichlet preconditioner. For a color version of this figure, see www.iste.co.uk/bouclier/IGA.zip*

As for the previous problems, we run the analysis for several levels of mesh refinement with and without the preconditioning step. The results of the study are given in Table 3.6. We also show the iterative convergence in Figure 3.14 for the refinement level equal to 2. We observe the same behavior as for the previous test cases. Indeed, the iteration count is drastically reduced with the use of a preconditioner, and it is all the more true when increasing the refinement level. We give the results for both the classical Dirichlet preconditioner $\tilde{\mathbf{F}}_I^{-1}$ and the generalized preconditioner $\tilde{\mathbf{F}}_A^{-1}$ in order to highlight the positive influence of the scaling step occurring in the second one. In fact, in the case of stiffened structures, the T-shape interfaces are sources of heterogeneities even if both sub-domains are of the same material properties. For example, the stiffener is mainly subjected to membrane forces, whereas the panel mostly undergoes bending. This leads to high differences in terms of stiffness at the interface. A shell is much stiffer under membrane loading than under bending loading. Thus, as shown in Table 3.6 and Figure 3.14, the generalized preconditioner performs better than the classical Dirichlet preconditioner for stiffened structures.

Once again, we mention that the proposed one-level FETI-based DD algorithm is not expected to be fully efficient in terms of iterative convergence when applied to fourth-order problems such as Kirchhoff–Love shells (see Farhat *et al.* (1998) for explanation in the context of standard FEM). However, the growth of the number of iterations on this quite complex test case remains slow when introducing the preconditioning step. In particular, satisfactory results are obtained even for very fine meshes. For example, with the refinement level of 4, the dual decomposition domain algorithm converges in 125 iterations even though the total number of DOF is over 100,000. This exhibits another crucial point of such solvers. Because the algorithms address an interface problem where the unknowns are the Lagrange multiplier DOF, we are able to measure, during the resolution, the residual of the kinematic coupling condition [3.28]. The convergence criteria are formulated on this residual, which, in other words, ensure the fulfillment of the coupling condition. For a large problem, the monolithic solution obtained with a basic direct solver applied directly to the linear system [3.25] may not give suitable results. The numerical errors during the resolution may significantly affect the correct imposition of the coupling conditions. Thus, we believe that the presented algorithm is viable for complex industrial applications because it enables us not only to quickly obtain a solution (parallel computing ability), but also to get an accurate solution on models, including massive amounts of DOF.

3.6. Summary and discussion

In order to design complex (such as industrial) structures in IGA, it is necessary to face the difficulty of handling multipatch models with possibly non-conforming parameterizations. A related objective of multipatch isogeometric analysis is to allow for a genuine link between CAD and analysis (and thus truly meets the initial motivation of IGA). This chapter tackles the problematic from an analysis point of view before the geometric modeling and optimization challenges are addressed in Chapter 4. A special care is taken to obtain a methodology that is also suitable for thin shell models since one of our intended applications is the design of aeronautic-type structures, i.e. made up of complex curved stiffened panels. To address this issue, we build in this chapter a family of dual DD solvers for the analysis of multipatch isogeometric solid elastic and Kirchhoff–Love shell models. The starting point is the implementation of a Mortar coupling scheme. If Mortar coupling in the field of standard solid elasticity is often encountered, the case of Kirchhoff–Love shells that can connect with arbitrary angles requires some developments. In this context, we propose a new Mortar coupling method that weakly imposes the interface conditions regarding the displacement and the rotation of the shell through two sets of Lagrange multipliers. In the literature, one argument against Mortar methods is commonly claimed; it concerns the additional unknowns introduced by the Lagrange multiplier field that increases the overall size of the system to be solved. We believe that this argument is not relevant and the present work fully takes advantage of these additional DOF. In fact, it allows us to simply formulate an interface problem, namely the one-level FETI problem, where the unknowns of the system are these interface DOF. We highlight for different examples with increasing levels of complexity that it enables us to accurately analyze multipatch structures (both using a standard solid or a non-conforming Kirchhoff–Love shell modeling), even for very fine levels of refinement, since we track the interface residual during the resolution. Moreover, the algorithms are naturally parallelizable. Instead of solving one large linear system, only local problems defined at the patch level are solved in parallel. As the non-invasive algorithm of the previous chapter, this is particularly of interest in the context of shape optimization since it enables us not to recompute the parts of the structure in non-design areas at each optimization iteration, as will be seen in Chapter 4.

Focusing on the proposed shell coupling approach, the extension to face real trimmed IGA (i.e. when some parts of the patches to be coupled are removed next to the interface) may be of great interest to enlarge the scope of applications and truly bridge the gap between CAD and shell analysis. One issue concerning the performance of such an extension might be related to the construction of suitable dual approximation spaces for the Lagrange multipliers (Wunderlich *et al.* 2019), which is not a trivial task, as underlined in the previous chapter. However, let us note that the case of real trimmed IGA differs from our case of non-conforming stiffened panels in terms of domains for integration but not in terms of basis functions involved in formulating the coupling conditions. As a result, our coupling approach may be relatively well adapted in order to truly handle trimmed IGA. Another solution would consist of developing an "equivalent" Nitsche method. As seen in Chapter 1 (see section 1.5.4), a connection between a standard Lagrange multiplier and (at least symmetric) Nitsche couplings can be made (see also Fritz *et al.* (2004); Bazilevs *et al.* (2012)). Starting with the standard Lagrange multiplier method, the idea to obtain the Nitsche method is to replace the Lagrange multipliers by the mean interface stress resultants coming from the primal variables. If some Nitsche strategies have been proposed for the case of G^1 junctions (Guo and Ruess 2015; Guo *et al.* 2018), the explicit formulation of the stress resultant transfer in the case of rigid hinges seems to be complicated. In this context, it may be mentioned that the very recent work of Schöllhammer and Fries (2019) regarding the development of a new Kirchhoff–Love shell formulation, which is not based on the curvilinear coordinates, suggests that it may be appealing to more simply write the stress resultant transfer occurring in Nitsche-type coupling.

Finally, returning to the DD solver, it is obvious that the true implementation in parallel would be of great interest to handle realistic complex industrial structures, as demonstrated in the field of FEM (Gosselet and Rey 2006). Regarding the improvement of the solver, one remaining question revolves around the choice of the operator $\mathbf{Q}$ in the coarse problem [3.41] to handle the heterogeneities (Rixen *et al.* 1999). Other interesting points to consider concern the extension of the FETI-DP framework (Farhat *et al.* 1998, 2001) and/or the construction of dedicated coarse spaces based on spectral techniques (Spillane and Rixen 2013; Spillane *et al.* 2013; Jolivet *et al.* 2014), or even the extension of the adaptive multipreconditioned FETI scheme (Bovet *et al.* 2017) to our case of non-conforming IGA, so as to be efficient with respect to both fourth-order problems and heterogeneities. The work presented in this chapter aims to motivate future research efforts in this direction. In our opinion, the direct use of geometric models with non-conforming interfaces for the structural analysis could be pushed forward with these new domain decomposition solvers.

4

Isogeometric Shape Optimization of Multipatch and Complex Structures

4.1. Introduction

Designing structural parts against the material limits, the impact of loads and many other constraints is a standard interest in engineering. However, finding the best possible design, namely the optimal design, is a complex discipline that combines mathematics and mechanics. A close link between the geometric and analysis model is essential in structural shape optimization since both models constantly communicate inside of the optimization loop. As mentioned in Chapter 1 (see, in particular, section 1.3.2), shape optimization is one of the greatest applications of IGA as it integrates CAD and analysis. This certainly explains why isogeometric structural shape optimization has been the object of intensive studies since the advent of IGA; see, for example, Wall *et al.* (2008), Qian (2010), Nagy *et al.* (2010), Nagy *et al.* (2011), Nagy *et al.* (2013), Kiendl *et al.* (2014), Taheri and Hassani (2014), Wang and Turteltaub (2015), Fußeder *et al.* (2015), Kang and Youn (2016), Lian *et al.* (2017), Wang *et al.* (2017a, 2018), Choi and Cho (2018), Lei *et al.* (2018), Hirschler *et al.* (2019b, 2020) and Weeger *et al.* (2019). A general procedure, which has been improved over the years (Daxini and Prajapati 2017; Wang *et al.* 2018), is commonly adopted. Roughly speaking, it is based on (i) a multilevel spline concept between the design and analysis models (Nagy *et al.* 2013; Kiendl *et al.* 2014; Wang and Turteltaub 2015; Hirschler *et al.* 2019b) and (ii) updating the shape by altering the spatial location of the control points.

From this overview, it appears that isogeometric shape optimization has not yet been truly applied to real-world structures. More precisely, simple geometries modeled as single-patch structures have mainly been considered. However, as underlined in the previous chapters of this book (see, for example, section 1.4.2), it is well known that a single spline patch cannot represent large and complex geometries. Here, we are interested in performing the shape optimization of complex multipatch structures with non-conforming interfaces. For the development of our method, we start by considering the case of shell structures with a particular emphasis on stiffened thin structures, which are omnipresent in aeronautics (see Chapter 3). The goal is to be able to properly optimize the shape and position of the stiffeners, and, if desired, the global shape of the panel. Such an issue has been faced in the standard FEM community: see, for example, Mulani *et al.* (2013), Locatelli *et al.* (2014) and Singh *et al.* (2017). It can be noted that these frameworks lack compactness as several softwares are used during the different steps of optimization (geometry parameterization, meshing, analysis), which necessarily involve some approximations, and thus severely affects the robustness and flexibility of the shape design process. The methodology is then applied and extended to other types of structures, such as general multipatch shells and solids. Numerous numerical results are provided to appreciate the performance of the proposed method.

More precisely, the developed framework begins with a careful integration of the DD solvers built in Chapter 3 in the shape optimization loop to achieve considerable computational time savings. Then, the key issue concerns the "geometrical" handling of non-conforming multipatch discretizations for shape optimization. For this purpose, we present a methodology inspired by free-form deformation (FFD) techniques (Sederberg and Parry 1986; Lyu and Martins 2015; Bletzinger 2017) and introduce embedded shell and solid formulations. It enables us to impose complex geometric constraints between non-conforming patches in a simple and suitable manner. Finally, we open the door for the efficient design of structural details within such complex macro-structures, by running a shape optimization process that calls upon the non-conforming non-invasive algorithm of Chapter 2. The study in this chapter is based on concepts introduced previously the previous chapters, in particular the general discussion in Chapter 1 regarding the ability of splines for shape modification and the remaining challenges within IGA for a true CAD-analysis bridge.

4.2. Isogeometric shape optimization framework

To start with, we outline the basics of structural isogeometric shape optimization by highlighting its main theoretical and practical points. In particular, we discuss the multilevel optimization process and the definition of the design variations. We also provide illustrations, with an emphasis on shells, since one of our intended applications concerns the optimal design of aerostructures. For further information on isogeometric shape optimization, the reader is advised to consult the references cited therein.

4.2.1. *Optimization flowchart*

Figure 4.1 describes the main steps of the shape optimization process. It has an initial CAD model as a main input. Here, the CAD model has to be understood as an analysis-suitable spline model, in the sense that it can be directly taken as input for numerical simulation (possibly using advanced numerical tools, such as the ones developed in Chapters 2 and 3; see also the discussion in section 1.4). As an output, the optimal shape is provided and, interestingly, this shape is described in terms of splines too. Therefore, the result is easily usable by designers and, thus, for production. There is no need to re-design the final geometry in a CAD environment as would be the case with FE-based optimization. From a global point of view, the optimal spline geometry is obtained from an iterative process with three steps. Each iteration consists of (see Figure 4.1):

1) updating the current shape into, hopefully, a better one;

2) running the analysis for this new geometry and then inferring mechanical and/or geometrical properties of interest;

3) computing new design variations to further improve the structure.

A more detailed description of each step of the process is given in the following paragraphs.

4.2.2. *Multilevel design*

A major asset of isogeometric shape optimization is the possibility of building, in a versatile manner, the design model (i.e. the geometrical model where the design variables allowing for shape modification are defined) and the analysis model (i.e. the numerical model used for the mechanical simulation). Indeed, benefiting from the flexible spline refinement procedures that do not alter the shape and regularity of the initial geometry (see section 1.2.3 and, for example, Cottrell *et al.* (2007)), both design and analysis models can be obtained through different refinements of the initial CAD model (see Figure 4.2). The discretization level of the associated design space depends on the complexity of the admitted shape variations. It is usually rather coarse to get "realistic" (smooth) shapes. In contrast, the analysis model is finer so as to ensure the good quality of the computations. Finally, thanks to the matrix representation of the refinement (see equations [1.25], [1.30] and [1.31]), the link between the two models is easily kept and, obviously, the two models represent the exact same geometry at each iteration of the optimization process (see Figure 4.1). In the field, we use the term *multilevel design* to denote this approach. It constitutes the main weapon that makes IGA extremely attractive for shape optimization (Nagy *et al.* 2010, 2011, 2013; Qian 2010; Nguyen *et al.* 2012; Kiendl *et al.* 2014; Fußeder *et al.* 2015; Wang and Turteltaub 2015; Herrema *et al.* 2017; Wang *et al.* 2017a; Hirschler *et al.* 2019b).

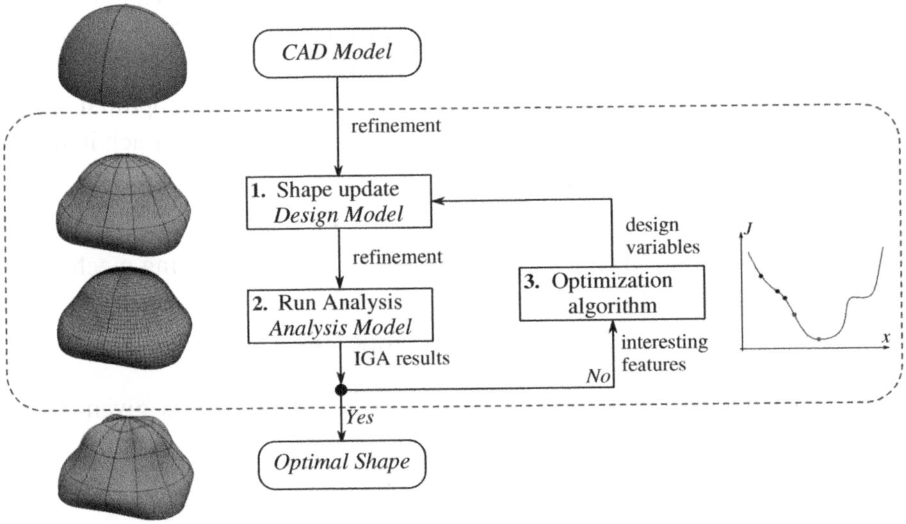

Figure 4.1. *Isogeometric shape optimization flowchart: overview of the main steps of the process. For a color version of this figure, see www.iste.co.uk/bouclier/IGA.zip*

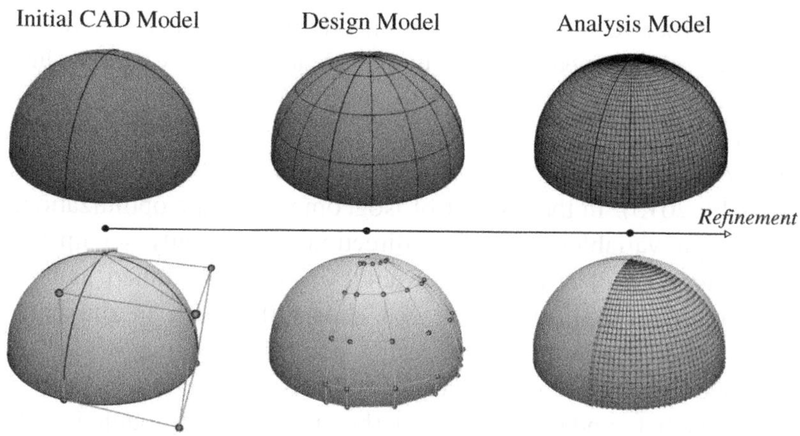

Figure 4.2. *Multilevel design approach: design and analysis spaces describe the exact same geometry and are initially obtained through different refinement levels of the initial CAD model: (top) spline (here NURBS) elements; (bottom) corresponding control points). For a color version of this figure, see www.iste.co.uk/bouclier/IGA.zip*

4.2.3. *Design variables*

The isogeometric framework also provides a proper and accurate way to vary the geometry. It has long been established that the shape control of splines is very comfortable and leads to smooth optimal design (Hasan Imam 1982; Braibant and Fleury 1984; Ramm *et al.* 1993). Indeed, as illustrated in section 1.2.3 (see Figure 1.12), a straightforward and natural way of modifying the shape of the structure is to move the control points. NURBS also offer the opportunity to apply shape variation by modifying the control point weights (see Figure 1.7), but this point is not investigated in this book. Indeed, unless a very coarse NURBS design model is considered (Nagy *et al.* 2010, 2013; Qian 2010), it appears that it is generally sufficient to only use the control point coordinates (in particular, for free-form surfaces, see, for example, Kiendl *et al.* (2014)).

Although attractive, taking the control point coordinates as design variables still requires some attention. The problems are similar to those encountered with FE-based optimization, where the FE nodal points are directly used as design variables. In fact, attention must be paid to mesh distortion in order to avoid local non-injectivities due to fold-overs. If the control points are allowed to move independently in every spatial direction, then the optimization process constantly needs to check the mesh quality and often requires cumbersome mesh regularization techniques (Bletzinger *et al.* 2010; Le *et al.* 2011; Nguyen *et al.* 2012; Firl *et al.* 2013; Bletzinger 2014; Choi and Cho 2015). In the context of isogeometric shape optimization, links between design variables and move directions are usually set up to avoid these challenges (Wall *et al.* 2008; Kiendl *et al.* 2014; Taheri and Hassani 2014; Wang *et al.* 2017a). Figure 4.3 shows this idea in the case of a hemisphere. By only letting each design control point move in the radial direction, we avoid mesh regularity problems. Moreover, control points located on patch boundaries are often (but not always) coupled with their neighbors to ensure a G^1 geometric continuity between patches (Kiendl *et al.* 2009). G^1-continuity means that two surfaces joining at a common edge have a common tangent plane at each point along that edge. This prevents the emergence of kinks during the optimization and it finally leads to smoother optimal shapes. Linking design variables may also be relevant in preserving initial shape properties, such as reflectional, rotational and translational symmetries. It also guarantees a reduction in the number of design variables that lead to lower computational costs and again, smoother optimal shapes.

4.2.4. *Formulation and resolution*

4.2.4.1. *A constrained optimization problem*

The mathematical formulation of an optimization problem involves different quantities. The objective function f quantifies the performance of the studied system. It is usually formulated so that the best solution is the one that returns the smallest value assessed by function f. The objective depends on several characteristics of the system, called variables or unknowns. In the specific case of structural optimization, we set design parameters that vary the geometry of the structure (as described in section 4.2.3). We denote these parameters as design variables and we represent them as a vector of unknowns s. In this work, we only consider the case of continuum variables. Finally, we describe the space in which we are looking for the optimal solution, through a combination of constraints c_i. Mathematically, the constraints are defined as scalar functions of s, which commonly take the form of implicit equations or inequalities. With these notations in hand, an optimization problem can be formulated as follows:

$$\begin{aligned} \text{minimize } & f(\mathbf{s}) \\ \text{w.r.t. } & \mathbf{s} \in \mathbb{R}^{n_s} \\ \text{subjected to } & c_i(\mathbf{s}) = 0, \ \forall i \in \mathcal{E} \\ & c_i(\mathbf{s}) \leq 0, \ \forall i \in \mathcal{I} \end{aligned} \tag{4.1}$$

where $\mathcal{E}$ and $\mathcal{I}$ are the set of indices for equality and inequality constraints, respectively.

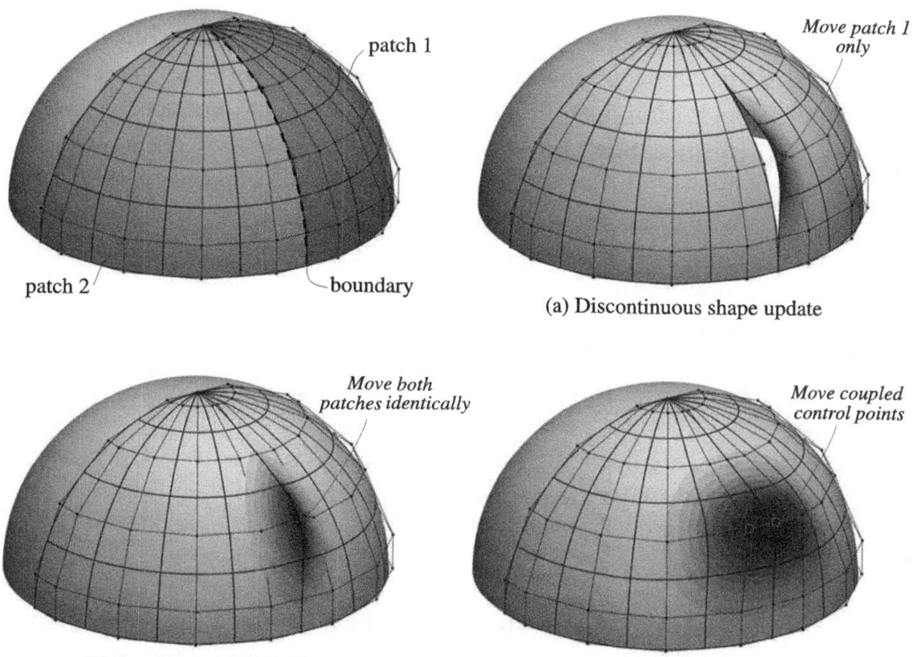

(a) Discontinuous shape update

(b) Continuous shape update

(c) G1-continuous shape update

Figure 4.3. *Shape modification of a hemisphere. The control points move radially. (a) However, the shape update can deteriorate the geometric continuity at the patch boundaries. (b) Boundary control points of the patches can be coupled to ensure a continuous geometry. (c) Boundary control points can be coupled with their neighbors to ensure G^1-continuity between patches. For a color version of this figure, see www.iste.co.uk/bouclier/IGA.zip*

In structural optimization, the goal is to improve the mechanical behavior of the structure. Thus, the expression of the objective function and/or the constraints involve quantities that describe the behavior of the structure. There can be various kinds of response functions. Typical quantities of interest for structural optimization are:

– the mass;

– the maximal deflection;

– the maximal stress;

– natural frequencies;

– the compliance;

– critical buckling loads

and surely others, depending on the design criteria. All of these functions can be expressed as explicit functions of the design variables and of the displacement field. Moreover, in the context of computational mechanics, the analysis is performed through an approximated method, such as IGA, for example. In this work, we consider the discrete approach for shape optimization, which means that the discretization step occurs before the formulation of the optimization problem (van Keulen *et al.* 2005). In this case, the objective function is expressed using the state variables $\mathbf{u}$ (i.e. the displacement DOF vector):

$$f := f\big(\mathrm{s}, \mathbf{u}(\mathrm{s})\big). \qquad [4.2]$$

In the case of static linear analysis, the variables $\mathbf{u}$ obviously depend on the design variables s through the linear system of equations:

$$\mathbf{K}(\mathrm{s})\mathbf{u} = \mathbf{f}(\mathrm{s}). \qquad [4.3]$$

Indeed, the stiffness matrix $\mathbf{K}$ and the load vector $\mathbf{f}$ are built through domain integrals. It means that their expressions depend on the shape of the structure, and consequently, on the design variables s. Finally, let us mention that the governing equations [4.3] only give an implicit relation between the state variables $\mathbf{u}$ and the design variables s. Thus, their dependency is not straightforward, in general.

Finally, let us more precisely describe the optimization problem that will be considered in the following. It probably consists of the most commonly adopted problem in structural design. It takes the compliance as the objective function and a given maximal volume (or, equivalently, the surface area in the case of shape optimization of shells with constant thickness) as a constraint. In other words, the goal is to maximize the overall rigidity of the structure under a certain limit of mass (volume and mass are proportional in the case of constant mass density). Additionally, the design space is limited by lower bounds $\mathbf{s}_l$ and upper bounds $\mathbf{s}_u$, which surround the n_s design variables. The optimization problem formally reads as:

$$\begin{aligned} \text{minimize } & f_{\text{cpl}}(\mathbf{s}, \mathbf{u}(\mathbf{s})) = \frac{1}{2}\mathbf{f}(\mathbf{s}) \cdot \mathbf{u}(\mathbf{s}) \\ \text{w.r.t. } & \mathbf{s} \in \mathbb{R}^{n_s} \\ \text{subjected to } & V(\mathbf{s})/V_0 - 1 \leq 0 \\ & \mathbf{s}_l \leq \mathbf{s} \leq \mathbf{s}_u \end{aligned} \qquad [4.4]$$

where $\cdot$ denotes the scalar product between two vectors and V_0 stands for the prescribed maximal volume.

4.2.4.2. *Gradient-based optimization algorithm*

Plenty of efficient numerical methods exist that enable us to solve the constrained optimization problems of form [4.4] (see, for example, Nocedal and Wright (2006)). Each algorithm is usually designed into a specific framework. In the case of structural shape optimization, *gradient-based* algorithms are often used. This is possible when the objective and constraint functions are differentiable with respect to the design variables. For large problems with hundreds to thousands of design variables, the use of gradients is quasi-inevitable to build an algorithm that converges in an acceptable amount of time. The information brought back by the gradients enables the algorithm to make suitable decisions during the resolution (Papadrakakis *et al.* 2001). For large-scale problems, gradient-free algorithms may require much more evaluation of the objective and constraint functions, which drastically increase the overall computational time.

More precisely, the optimization problem is solved iteratively: at a given iteration k, the design variables are updated by computing a correction term $\mathbf{\Delta}^{(k)} \in \mathbb{R}^{n_s}$, such that the new iterate reads:

$$\mathbf{s}^{(k+1)} = \mathbf{s}^{(k)} + \mathbf{\Delta}^{(k)}. \qquad [4.5]$$

In the case of gradient-based algorithms, the correction term depends on the gradient of the objective functions (and of the constraints), with respect to the design variables computed at the current iterate:

$$\boldsymbol{\Delta}^{(k)} := \boldsymbol{\Delta}\left(\frac{df}{ds}(\mathbf{s}^{(k)})\right). \tag{4.6}$$

Even more so, standard algorithms often recycle the gradients from the previous iterations, in particular to approximate the Hessian of the objective function:

$$\boldsymbol{\Delta}^{(k)} := \boldsymbol{\Delta}\left(\frac{df}{ds}(\mathbf{s}^{(k)}), \frac{df}{ds}(\mathbf{s}^{(k-1)}), \ldots, \frac{df}{ds}(\mathbf{s}^{(0)})\right). \tag{4.7}$$

This work is not focused on the development of a new, specific optimization algorithm. Here, we employ the gradient-based algorithm `SLSQP` (Kraft 1988) available in many scientific packages, as is, for instance, `SciPy` (Jones *et al.* 2001). From the basic numerical procedures, this algorithm is based on the SQP method (Fletcher 1987), which can be viewed as an extension of the Newton method for constrained optimization. At each major iteration, an approximation is made on the Hessian of the Lagrangian function associated with [4.4], using a quasi-Newton updating method. These methods are appealing because the Hessian is not directly computed, but approximated through the gradient variations during the resolution, thereby offering simplicity, minimal computational cost and good convergence properties. A popular method for this approximation is the BFGS Hessian update strategy (Broyden 1970).

NOTE.– Another important issue in general gradient-based optimization concerns the initialization of the algorithms, since these only provide a critical point, i.e. such that the gradient of the functional vanishes, which may not be a global minimum in the case of non-convex problems (Nocedal and Wright 2006). For structural shape optimization, the initial guess is straightforward: this is the first shape provided by the designers according to their experiences. Thus, the process of Figure 4.1 modifies the shape of the structure in the neighborhood of the initial geometry, which, most of time, results in a mechanical response, significantly improved from that of the initial configuration. Finally, note that the multilevel design concept outlined in section 4.2.2 can be easily extended to improve the performance of the optimization procedure: starting with a very coarse discretization, the design model can be refined successively each time a minimum is reached, which results in an efficient filtering strategy, as shown by Bandara and Cirak (2017).

4.2.4.3. *Sensitivity analysis*

Ineluctably, the crucial issue when using gradient-based optimization algorithms is the computation of the sensitivities, i.e. the gradient of the response functions with respect to the design variables.

A simple way to perform the sensitivity analysis is to resort to global finite difference, which consists of perturbing each design variable with a small variation and evaluating the response of the structure for each of these perturbations. The use of the finite difference (FD) formula enables us to then obtain the derivatives. The most popular remains the forward difference:

$$\frac{df}{ds_i}(\mathbf{s}) \approx \frac{f(\mathbf{s}+\varepsilon\,\mathbf{e}_i) - f(\mathbf{s})}{\varepsilon}, \qquad [4.8]$$

where the perturbation ε takes a small scalar value, for example, $\varepsilon = 10^{-6}$. Vector $\mathbf{e}_i$ has the same length as s, with all components equal to zero, except component i, which is set to 1. The accuracy depends on the parameter ε. In theory, the exact derivative is obtained when it tends to zero. However, in practice, the step size is limited due to arithmetic accuracy and a value that is too small gives the wrong results. Furthermore, if the function f depends on the state variables (as is the case for the compliance), then the state equations [4.3] have to be solved for each perturbation, which turns out to be very computationally demanding in the case of a large number of design variables.

Alternatives to global finite difference sensitivities exist, namely semi- or even full analytical derivation. Indeed, we can apply a chain rule of differentiation starting with equation [4.2]. The total derivative of function f *w.r.t.* a design variable s_i reads:

$$\frac{df}{ds_i} = \frac{\partial f}{\partial s_i} + \frac{\partial f}{\partial \mathbf{u}} \cdot \frac{d\mathbf{u}}{ds_i}. \qquad [4.9]$$

The term $d\mathbf{u}/ds_i$ is not explicitly known. We can identify this term by differentiating the discrete state equations [4.3]. It yields:

$$\frac{d\mathbf{u}}{ds_i} = \mathbf{K}^{-1}\left(\frac{\partial \mathbf{f}}{\partial s_i} - \frac{\partial \mathbf{K}}{\partial s_i}\mathbf{u}\right). \qquad [4.10]$$

Then, we can substitute equation [4.10] into equation [4.9]:

$$\frac{df}{ds_i} = \frac{\partial f}{\partial s_i} + \frac{\partial f}{\partial \mathbf{u}} \cdot \mathbf{K}^{-1} \left(\frac{\partial \mathbf{f}}{\partial s_i} - \frac{\partial \mathbf{K}}{\partial s_i} \mathbf{u} \right). \quad [4.11]$$

One can see that the inverse $\mathbf{K}^{-1}$ is involved in the expression of the derivative. To a certain extent, it means that a resolution is required. There are two ways of dealing with this issue, namely the adjoint and direct approaches. In the adjoint method, we first perform the product $\frac{\partial f}{\partial \mathbf{u}} \cdot \mathbf{K}^{-1}$, i.e. we solve the adjoint problem:

$$\mathbf{K}\mathbf{u}^{\mathrm{ad}} = \frac{\partial f}{\partial \mathbf{u}}. \quad [4.12]$$

The adjoint solution does not depend on the design variables. Consequently, for each design variable, the adjoint solution $\mathbf{u}^{\mathrm{ad}}$ is reused. Finally, the complete expression of the total derivative reads as follows:

$$\frac{df}{ds_i} = \frac{\partial f}{\partial s_i} + \mathbf{u}^{\mathrm{ad}} \cdot \left(\frac{\partial \mathbf{f}}{\partial s_i} - \frac{\partial \mathbf{K}}{\partial s_i} \mathbf{u} \right). \quad [4.13]$$

Conversely, the direct approach consists of starting with the product:

$$\mathbf{K}^{-1} \left(\frac{\partial \mathbf{f}}{\partial s_i} - \frac{\partial \mathbf{K}}{\partial s_i} \mathbf{u} \right),$$

i.e. in solving several linear systems with the so-called pseudo-load vectors as right-hand sides:

$$\mathbf{K}\mathbf{u}_i^{\mathrm{dir}} = \frac{\partial \mathbf{f}}{\partial s_i} - \frac{\partial \mathbf{K}}{\partial s_i} \mathbf{u}. \quad [4.14]$$

Once the solution $\mathbf{u}_i^{\mathrm{dir}}$ is computed, we can obtain the total derivative:

$$\frac{df}{ds_i} = \frac{\partial f}{\partial s_i} + \frac{\partial f}{\partial \mathbf{u}} \cdot \mathbf{u}_i^{\mathrm{dir}}. \quad [4.15]$$

Thus, in the direct sensitivity analysis, there are potentially as many systems to be solved as design variables. Therefore, adjoint sensitivity analysis is often preferred, as long as the number of response functions involved in the optimization problems (objectives and constraints) is smaller than the number of design variables.

Let us now take our case of the structural compliance as a response function (see equation [4.4]). We can specify the expression of the different terms involved in the general formulation of the derivative [4.11]:

$$\frac{\partial f_{\text{cpl}}}{\partial s_i} = \frac{1}{2}\frac{\partial \mathbf{f}}{\partial s_i} \cdot \mathbf{u}, \qquad \text{and} \qquad \frac{\partial f_{\text{cpl}}}{\partial \mathbf{u}} = \frac{1}{2}\mathbf{f}. \qquad [4.16]$$

In this case, the problem is self-adjoint. In fact, the adjoint problem reads here as:

$$\mathbf{K}\mathbf{u}^{\text{ad}} = \frac{1}{2}\mathbf{f} \quad \Rightarrow \quad \mathbf{u}^{\text{ad}} = \frac{1}{2}\mathbf{u}. \qquad [4.17]$$

Thus, the adjoint solution is deduced from the state solution and no additional resolution is needed. Finally, with these developments in hand, we can obtain the following derivative of the compliance:

$$\frac{df_{\text{cpl}}}{ds_i} = \mathbf{u} \cdot \left(\frac{\partial \mathbf{f}}{\partial s_i} - \frac{1}{2}\frac{\partial \mathbf{K}}{\partial s_i}\mathbf{u} \right). \qquad [4.18]$$

The very last point consists of the computation of the derivative of the stiffness matrix and the load vector, with respect to the design variables. This last step is the most challenging. Their expressions depend on the nature of the design variables and also on the finite element formulation.

In semi-analytical discrete sensitivities, the calculation of the derivatives of the stiffness matrix and of the load vector, with respect to the design variables, are simplified by approximating them using FD (same process as in [4.8]):

$$\frac{\partial \mathbf{f}}{\partial s_i} \approx \frac{\Delta \mathbf{f}}{\Delta s_i} = \frac{\mathbf{f}(\mathbf{s} + \varepsilon\, \mathbf{e}_i) - \mathbf{f}(\mathbf{s})}{\varepsilon}, \qquad [4.19]$$

$$\frac{\partial \mathbf{K}}{\partial s_i} \approx \frac{\Delta \mathbf{K}}{\Delta s_i} = \frac{\mathbf{K}(\mathbf{s} + \varepsilon\, \mathbf{e}_i) - \mathbf{K}(\mathbf{s})}{\varepsilon}. \qquad [4.20]$$

With this strategy, we have to build as many stiffness matrices as the number of design variables. However, in contrast to global FD sensitivities [4.8], there is no need to solve the state governing equations for each design variable. Moreover, the stiffness matrix is defined as a sum of elementary matrices of a smaller size. One single design variable may not be related to every elementary matrix. Thus, the stiffness variations are preferably approximated at the elementary level. For the compliance, it gives something of the form:

$$\frac{df_{\text{cpl}}}{ds_i} \approx \sum_{e \in \mathcal{F}_i} \mathbf{u}^e \cdot \left(\frac{\Delta \mathbf{f}^e}{\Delta s_i} - \frac{1}{2} \frac{\Delta \mathbf{K}^e}{\Delta s_i} \mathbf{u}^e \right), \qquad [4.21]$$

where $\mathcal{F}_i$ is the set of active elements corresponding to the design variable s_i. One interesting feature of semi-analytical sensitivities, besides the appearing simplicity, is the independence to the element formulation. This is very convenient from an implementation point of view. No element dependent routine has to be formulated. Therefore, an optimization code including semi-analytical sensitivities can be applied to any finite element formulation.

Nevertheless, it has been shown that semi-analytical sensitivities suffer from a very specific type of approximation error (Olhoff *et al.* 1993; Bletzinger *et al.* 2008; Firl 2010; Wang *et al.* 2015; Bletzinger 2017). In fact, they lead to inconsistent derivatives regarding a rigid-body test, as discussed in Bletzinger (2017). Some methods have been suggested to correct the derivatives, see, for example, Olhoff *et al.* (1993), Bletzinger *et al.* (2008) and Wang *et al.* (2015). Without any correction, semi-analytical derivatives are sensitive to the mesh size and step size. Finally, let us mention that the computational cost of these sensitivities is still proportional to the number of design variables. Thankfully, within the isogeometric framework, it is possible to go one step further and compute the sensitivities in a full analytical fashion. For that purpose, we follow the chain rule of differentiation, in order to go from the definition of the design variables to the analysis model, where the finite element operators are formulated. Basically, sensitivities are viewed as:

$$\frac{df}{ds_i} = \frac{\partial f}{\partial s_i} + \frac{d\mathbf{x}}{ds_i} \frac{d\hat{\mathbf{x}}}{d\mathbf{x}} \left(\frac{\partial \mathrm{W}_{\text{int}}}{\partial \hat{\mathbf{x}}} + \frac{\partial \mathrm{W}_{\text{ext}}}{\partial \hat{\mathbf{x}}} \right), \qquad [4.22]$$

where:

– $d\mathbf{x}/ds_i$ is the differentiation of the shape parameterization (link between the design variables and the control point locations $\mathbf{x}$ of the design model);

– $d\hat{\mathbf{x}}/d\mathbf{x}$ is the differentiation of the refinement step occurring between the design model (associated control points $\mathbf{x}$) and the analysis model (corresponding control points $\hat{\mathbf{x}}$), which consists of a linear operation (see, for example, [1.31]);

– $\partial \mathrm{W}_{\text{int}}/\partial \hat{\mathbf{x}}$ and $\partial \mathrm{W}_{\text{ext}}/\partial \hat{\mathbf{x}}$ are the partial differentiation of the internal and external works computed at the level of the analysis model, respectively.

Due to the tight link between all of the involved quantities in the IGA framework, those derivatives are achievable without many difficulties. Contrary to semi-analytical sensitivities, no approximations are introduced here. But, more interestingly, these analytical sensitivities have a computational cost that is (quasi) independent of the number of design variables. It has been observed that the computational costs of these analytical sensitivities are similar to the cost required to evaluate the internal and external works themselves, which is considerably small in comparison with approximated sensitivities. We do not go into further detail here; the reader is advised to consult Qian (2010); Nagy *et al.* (2013); Taheri and Hassani (2014); Lei *et al.* (2018) and Hirschler *et al.* (2020) for further insight regarding the full expressions of analytical sensitivities. Especially, in Hirschler *et al.* (2020) all of the formulas and several algorithms are given to implement the aforementioned sensitivity analysis.

4.3. Unify the DD approach and multipatch optimization: towards ultimate efficiency

Structural optimization is computationally expensive because it requires a large successive number of structural (and sensitivity) analyses (see Figure 4.1). If the goal is to design complex, real-world structures, then it is crucial that these computations are efficiently performed. In this context, the use of the non-conforming DD solvers developed in Chapter 3 is highly relevant, since they allow for a parallel computation of the solution of the state equations [4.3] (see also step 2 in Figure 4.1) and the parallel architecture is adapted to compute the response functions and the sensitivities, and computational savings can be easily met in the case of non-design areas and smart re-analysis.

4.3.1. *DD computation of the response functions*

Standard response functions, such as those mentioned in section 4.2.4, can be computed in a parallel way. Indeed, in the case of multipatch structures (made of multiple parts), the overall response function can usually be expressed as the sum of local, patch-wise quantities, similarly to the matrix–vector product involving the dual Schur complement of the state problem (see equation [3.42]). For example, in the case of the compliance, we can write:

$$f_{\text{cpl}} = \left(f_{\text{cpl}}\right)^{\text{part 1}} + \left(f_{\text{cpl}}\right)^{\text{part 2}} + \cdots + \left(f_{\text{cpl}}\right)^{\text{part n}}, \qquad [4.23]$$

where $\left(f_{\text{cpl}}\right)^{\text{part } k}$ is the compliance computed over the sub-domain k, i.e. by making use of the notations in Chapter 3 (see, in particular, equation [3.27]):

$$\left(f_{\text{cpl}}\right)^{\text{part } k} = \frac{1}{2}\left(\mathbf{f}_k - \mathbf{C}_k^{F^T} \boldsymbol{\lambda}\right) \cdot \mathbf{u}_k. \qquad [4.24]$$

Therefore, once the solution of the state problem is obtained using the DD solvers in Chapter 3, the compliance can be computed using the same parallel architecture, i.e. by computing independently local contributions and then gathering all of them through a sum.

4.3.2. *DD computation of the sensitivities*

Even more interestingly than the response functions, the sensibilities can also be computed in a parallel way by exploiting the decomposition of the computational domain.

First, except for the case of the compliance that leads to a self-adjoint problem, the calculation of the sensitivities requires the resolution of a system (or even multiple). Either for the direct method [4.15] or for the adjoint method [4.13], the systems to be solved involve the same operators as in the state problem (but different right-hand sides; see equations [4.14] and [4.12], respectively). As a result, the DD algorithm can surely be used to solve the linear systems involved during the sensitivity analysis. Consequently, the benefits highlighted in Chapter 3 for the computation of $\mathbf{u}$ also apply for the computation of the sensitivities. Furthermore, since the pseudo-inverses $\mathbf{K}_s^+$ have already been built to perform the structural analysis, the numerical cost for solving the linear systems during the sensitivity analysis can be largely reduced. In some sense, the local sub-problems are already solved.

Second, identically to equation [4.23], the computation of the sensitivities can be done by gathering local independent sensitivities. For example, the sensitivities of the compliance can be computed as a sum of local sensitivities:

$$\frac{df_{\text{cpl}}}{ds_i} = \left(\frac{df_{\text{cpl}}}{ds_i}\right)^{\text{part 1}} + \left(\frac{df_{\text{cpl}}}{ds_i}\right)^{\text{part 2}} + \cdots + \left(\frac{df_{\text{cpl}}}{ds_i}\right)^{\text{part n}} \qquad [4.25]$$

where the sensitivities related to the sub-domain k read:

$$\left(\frac{\partial f_{\text{cpl}}}{\partial s_i}\right)^{\text{part } k} = \mathbf{u}_k \cdot \left(\left(\frac{\partial \mathbf{f}_k}{\partial s_i} - \frac{\partial \mathbf{C}_k^{F^T}}{\partial s_i}\boldsymbol{\lambda}\right) - \frac{1}{2}\frac{\partial \mathbf{K}_k}{\partial s_i}\mathbf{u}_k\right). \qquad [4.26]$$

At this stage, let us point out that, in general, one sub-domain only contributes to a part of the full gradient, which means that equation [4.26] does not vanish for a limited set of design variables. Indeed, the local load vector $\mathbf{f}_k$, the coupling operator $\mathbf{C}_k^F$ and the local stiffness matrix $\mathbf{K}_k$ depend on the design variables, which act directly on the geometry of the current sub-domain k, and those which are associated with its neighboring sub-domains (more specifically, those that will modify the coupling interfaces). From an implementation point of view, the $n_{s,k}$ active design variables associated with sub-domain k can be extracted through a Boolean $n_{s,k} \times n_s$ matrix $\mathbf{B}_k$, such that:

$$\mathbf{s}_k = \mathbf{B}_k \mathbf{s}, \qquad [4.27]$$

where $\mathbf{s}_k$ collects the local design variables. Making use of this selection operator, the local contributions are finally gathered, as follows, to obtain the full gradient:

$$\frac{df_{\text{cpl}}}{d\mathbf{s}} = \sum_k \mathbf{B}_k^T \left(\frac{df_{\text{cpl}}}{d\mathbf{s}_k}\right)^{\text{part } k}, \qquad [4.28]$$

where the local gradients take the form of real-valued column vectors of size $n_{s,k}$.

4.3.3. *Non-design parts*

In the case of a complex structure built through the assembly of multiple sub-parts, the optimization problem may only involve the shape modification of some of these sub-parts. For example, this happens for stiffened structures: if the goal is to only optimize the shape and position of the stiffeners, then the main structure (i.e. the panel) remains unchanged during the optimization process. The panel is considered as a non-design part in this case.

Interestingly, the non-conforming DD solver enables us to naturally perform computational savings when there are non-design parts. Indeed, let us recall that DDMs are based on the separation of the sub-parts: the finite element operators are not globally assembled. Instead, only local subproblems are built and solved. More specifically, the sub-domain stiffness matrix associated with a non-design part $\mathbf{K}_s$ will not change during the whole optimization process. This is especially true with the use of non-conforming discretizations. Without the condition of matching meshes, we keep the meshes associated with the non-design parts unchanged (see Figure 1.21 as a reminder). As a result, the pseudo-inverses $\mathbf{K}_s^+$ associated with the non-design parts need to be built once and for all at the beginning of the resolution (of the optimization problem). This can represent great numerical savings.

4.3.4. *Fast re-analysis*

During the iterative resolution of a structural optimization problem, successive structural and sensitivity analyses are performed. Depending on the amount of shape changes from one design to another, these analysis results may not be too far from one another. Consequently, we may reuse the results of the previous optimization iterations to accelerate the analyses at the current iteration. For example, regarding the PCPG algorithm (see Table 3.1) involved in the dual DD solver, the choice of the starting point $\boldsymbol{\lambda}^{(0)}$ may be improved by simply taking the solution of the previous design.

Another interesting idea consists of recycling the previous factorization when solving the local Dirichlet subproblems involved in the preconditioner (remember equation [3.47]). Indeed, there may be no great benefit to exactly recompute the local primal Schur complements involved during the preconditioning for each optimization iteration. As long as the number of iterations required by the DD solver does not increase, we can keep using the same factorizations of the local stiffness matrices associated with the internal DOF (i.e. $\mathbf{K}_{s_{ii}}^{-1}$ in equation [3.47]) used during the previous optimization iterations. If the number of iterations required by the DD solver grows largely, this indicates that the preconditioning step is deteriorated. It is then the time to pay the price of recomputing these factorizations.

Taking the reasoning even further, we may finally apply the same idea for the computation of the local solution themselves, i.e. involving the pseudo-inverses $\mathbf{K}_s^+$ (see equation [3.42]). In this respect, we can imagine solving the local Neumann problems (step 2 in the figure below equation [3.42]) with a preconditioned iterative solver, where the associated preconditioner would be $\mathbf{K}_s^+$ of the previous optimization iterations, as long as the iteration counts of the DD solver are essentially constants. Such a process would lead to computational savings regarding the resolution of the local problems. Indeed, the building of the pseudo-inverses takes a non-neglected part of the overall computational cost (sometimes more than 50% through our experience).

4.3.5. *Optimization algorithm*

Finally, let us mention that the optimization algorithm itself could have been designed in accordance with the domain decomposition philosophy. Instead of solving one large optimization problem, we can potentially solve local optimization problems in addition to the coarse optimization problems that link them together. Such an approach would be closely related to the field of multidisciplinary optimization (Cramer *et al.* 1994; Balesdent *et al.* 2011; Martins and Lambe 2013a), and interested readers could surely find perspectives for future research there.

4.4. Innovative design of multipatch structures: focus on aeronautical stiffened panels

New methodologies become truly attractive when they are able to solve problems that are of real interest. Applying the isogeometric shape optimization strategy introduced in section 4.2 for designing large, complex (and thus multipatch) structures is a challenging task from both the geometric and analysis sides. From the analytical point of view, we have built an efficient strategy to couple non-conforming discretizations in Chapter 3, and have highlighted its deep interest when finely integrated into a shape optimization loop, above. The goal is now to address the geometrical challenge, i.e. to build an adequate methodology in order to properly impose the shape modification of complex isogeometric structures with multiple parts. As explained in the introduction of this chapter, here, we introduce the developed approach in the specific case of shells, while the application to solids is postponed to section 4.5. The methodology described in this section originates from Hirschler *et al.* (2019a). In particular, we will make use of such an approach to successfully design large stiffened aerostructures in section 4.6.

4.4.1. *Geometric modeling: embedded entities*

Our approach is based on the use of embedded (or immersed) entities to handle the geometric constraint of keeping the different parts of the structures connected during shape modification. The associated geometric modeling is introduced in the following.

4.4.1.1. *Geometric challenges*

To illustrate the geometric challenges to be faced, let us focus on the simple example of a portion of a cylinder, with a flat stiffener defined in either the circumferential or in the longitudinal direction, as depicted in Figure 4.4 (left). Two classical spline surfaces describe the portion of the cylinder and the stiffener. During the design, or more specifically, the optimization process, the designer may want to modify the shape of the different parts by moving the control points of the design spline model (following the procedure outlined in section 4.2 and Figure 4.1).

In the case of the circumferential stiffener, the control points associated with the stiffener can be moved, for example, along the longitudinal direction. This shape modification is illustrated in Figure 4.4 (top) and does not exhibit any particular problem. However, it becomes more tricky with the longitudinal stiffener. To modify its shape, we can try to move the control points into the circumferential direction, as described in Figure 4.4 (middle). The first difficulty arises. We want the bottom of the stiffener to lie perfectly on the cylinder. This geometric constraint is not easily fulfilled: sliding the control points along the edge of the cylinder disconnects the surfaces. The second difficulty is modifying the global shape (here, the portion of cylinder), while keeping the substructure connected (see Figure 4.4 (bottom)). More precisely, the question is: how do we transfer the shape update from the global part to its substructure? If we move one control point of the main surface, the stiffener does not lie on it anymore. In this simple case, we could formulate an appropriate condition in order to glue the stiffener, but, in general, it is not a trivial task. In the case of curvilinear stiffeners lying on a curved surface, re-parameterizations and approximations are inevitable if the standard isogeometric strategy for shape optimization is adopted. Instead, here, we are looking for a compact approach that does not require cumbersome re-parameterization procedures. This is particularly crucial to get a robust shape optimization strategy.

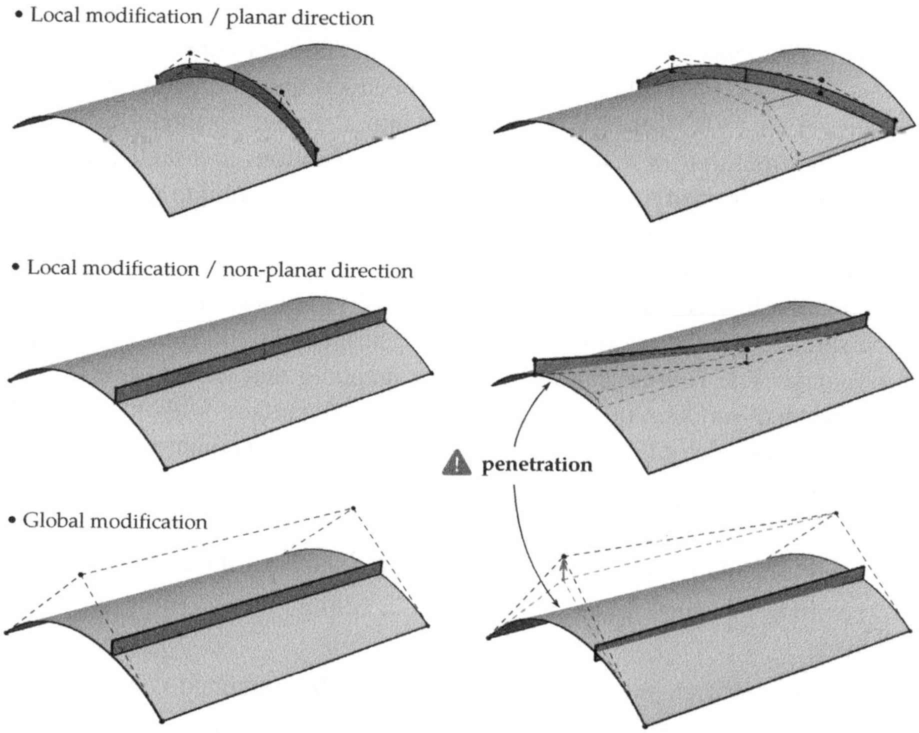

Figure 4.4. *Geometric difficulties to impose design modifications to a stiffened cylinder: (top) a movement of the stiffener along the longitudinal direction of the panel does not present any particular problem; however (middle), while moving in the circumferential direction, the stiffener disconnects with the cylinder, and (bottom) the stiffener does not follow the shape variation of the cylinder. For a color version of this figure, see www.iste.co.uk/bouclier/IGA.zip*

Recapitulating, we are looking for a geometric modeling that enables us to impose the following shape modifications in a convenient way:

– global shape optimization of a whole structure that contains multiple parts;

– local shape optimization of specific sub-parts (such as stiffeners).

4.4.1.2. *Using spline compositions for shape updating*

4.4.1.2.1. Embedded entities

We introduce the concept of embedded surfaces in order to overcome the geometric challenges highlighted in Figure 4.4. The idea comes from Bauer *et al.* (2017) who considered univariate embedded entities. More precisely, in Bauer *et al.* (2017), they embed curves into a surface to model cables in membrane structures, or to apply line boundary conditions to a shell structure, without being limited to the edges of the surface. A similar procedure is undertaken to define the sub-parts (see Figure 4.5 for an illustration in the case of a stiffened panel). We do not use classical spline surfaces to describe the sub-parts. Instead, we introduce the composition of a spline volume and a spline surface. In other words, we immerse a surface into the parameter space of a volume. The volume can be seen as a mapping that smartly transforms the embedded surface, in order to create a final surface with the desired properties. This idea offers a novel strategy to generate geometric models of thin structures, such as stiffened structures. The new approach is formalized in Figure 4.6. The construction contains four steps:

1) build the main part of the structure using a standard spline surface;

2) extrude this surface so as to generate a volume in which the sub-parts will live;

3) embed surfaces into the parameter space of the generated volume;

4) transform these embedded surfaces using the volume mapping.

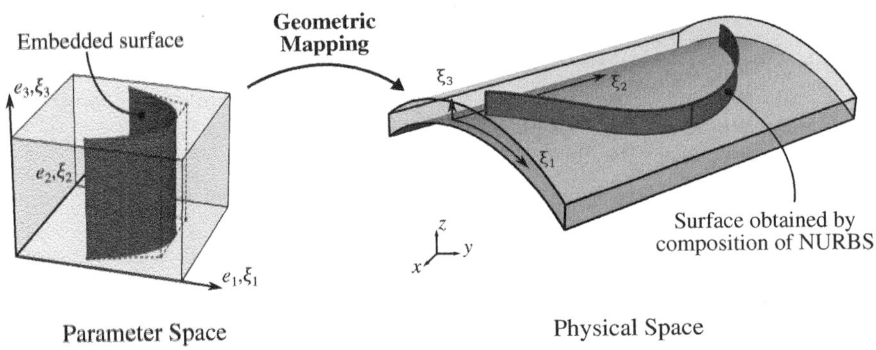

Figure 4.5. *Construction of a stiffened structure using embedded surfaces. The final surface describing the stiffener is obtained by the composition of a spline surface and a spline volume. This construction ensures that the stiffener lies perfectly on the portion of the cylinder. For a color version of this figure, see www.iste.co.uk/bouclier/IGA.zip*

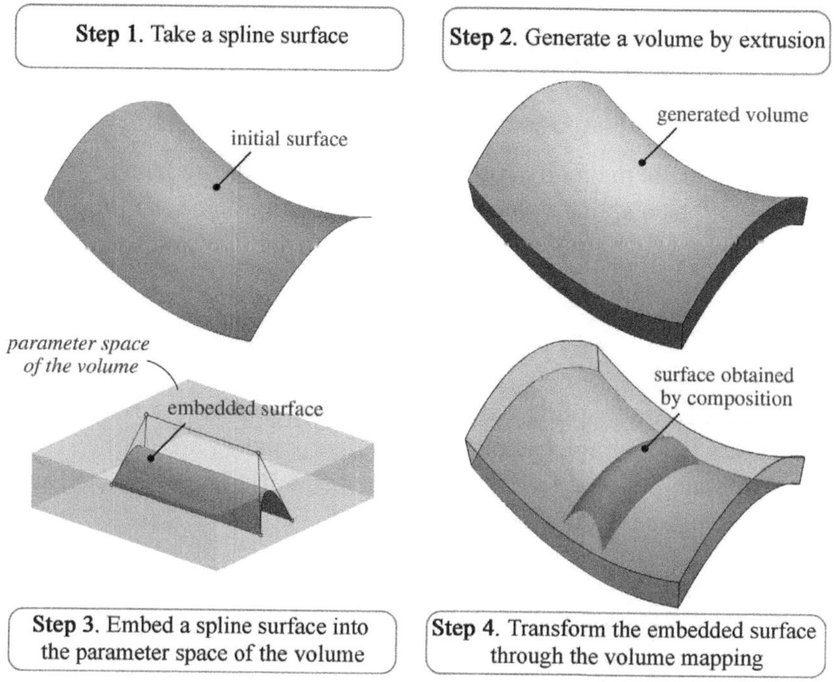

Figure 4.6. *Generating a geometric complex model of a thin structure by using the embedded strategy*

The proposed strategy can be related to free-form deformation (FFD) techniques, which are commonly used in graphic design to deform objects (Sederberg and Parry 1986) and are also applied, for example, for aerostructural optimization (Duvigneau 2006; Lassila and Rozza 2010; Kenway and Martins 2014; Lyu and Martins 2015). In that sense, the design approach presented here is not entirely disconnected from the current industrial methods. However, note in the following that the transformation of the full embedded entity will be performed, whereas only the FE-node locations are altered by the volume mapping in the standard FE-based FFD approach (Lyu and Martins 2015; Bletzinger 2017). From here on, we will interpret the embedded entities as *slave parts* as their shapes are driven by external geometric properties. Conversely, the *master parts* are those which dictate specific constraints during the shape update. Returning to the example depicted in Figure 4.5, the portion of cylinder is the master part and the stiffener constitutes the slave part: modifying the master part imposes the transfer of shape modification to the slave part and, in turn, the possible shape modifications of the slave part are restricted to follow the master geometry.

4.4.1.2.2. Mathematical description

Let us now mathematically formulate the geometric modeling. We define the mapping V^h as a spline volume, i.e. the following notations introduced in section 1.2:

$$V^h(\xi_1, \xi_2, \xi_3) = \sum_{i=1}^{n_{3D}} R_i(\xi_1, \xi_2, \xi_3) P_i, \qquad [4.29]$$

where R_i are trivariate spline basis functions and P_i are the related control point locations. The parametric coordinates ξ_1, ξ_2 and ξ_3 are defined over the parameter domain denoted by $\bar{\Omega}$ (see Figure 4.5, left). The embedded surface $\bar{S}^h$ is a spline surface defined as:

$$\bar{S}^h(\theta_1, \theta_2) = \sum_{j=1}^{n_{2D}} \bar{R}_j(\theta_1, \theta_2) \bar{P}_j, \qquad [4.30]$$

with $\bar{R}_j$ as some bivariate spline basis functions and $\bar{P}_j$ as the location of the associated control points living in the parameter space of the volume. As in section 3.3.1, θ_1 and θ_2 are the parametric coordinates associated with the surface S^h. Under the consideration that the codomain of $\bar{S}^h$ is included in domain $\bar{\Omega}$, we finally compose the two previous mappings [4.29] and [4.30] to get the physical surface:

$$S^h(\theta_1, \theta_2) = V^h \circ \bar{S}^h(\theta_1, \theta_2) = V^h\left(\ \bar{S}^h(\theta_1, \theta_2)\ \right). \qquad [4.31]$$

The final surface S^h is parameterized by the same set of variables (θ_1, θ_2) as the embedded surface $\bar{S}^h$ and it returns value in the physical space $\mathbb{R}^3$.

4.4.1.2.3. Potential for shape optimization

The potential of using embedded entities is multiple from the design point of view. Our first motivation was to design stiffened structures found in aeronautics, but it should absolutely not be limited to this purpose. Figure 4.7 shows the potential of the presented approach from the design point of view. If the approach eases the construction of complex structures, it also offers interesting features related to shape optimization in a general context.

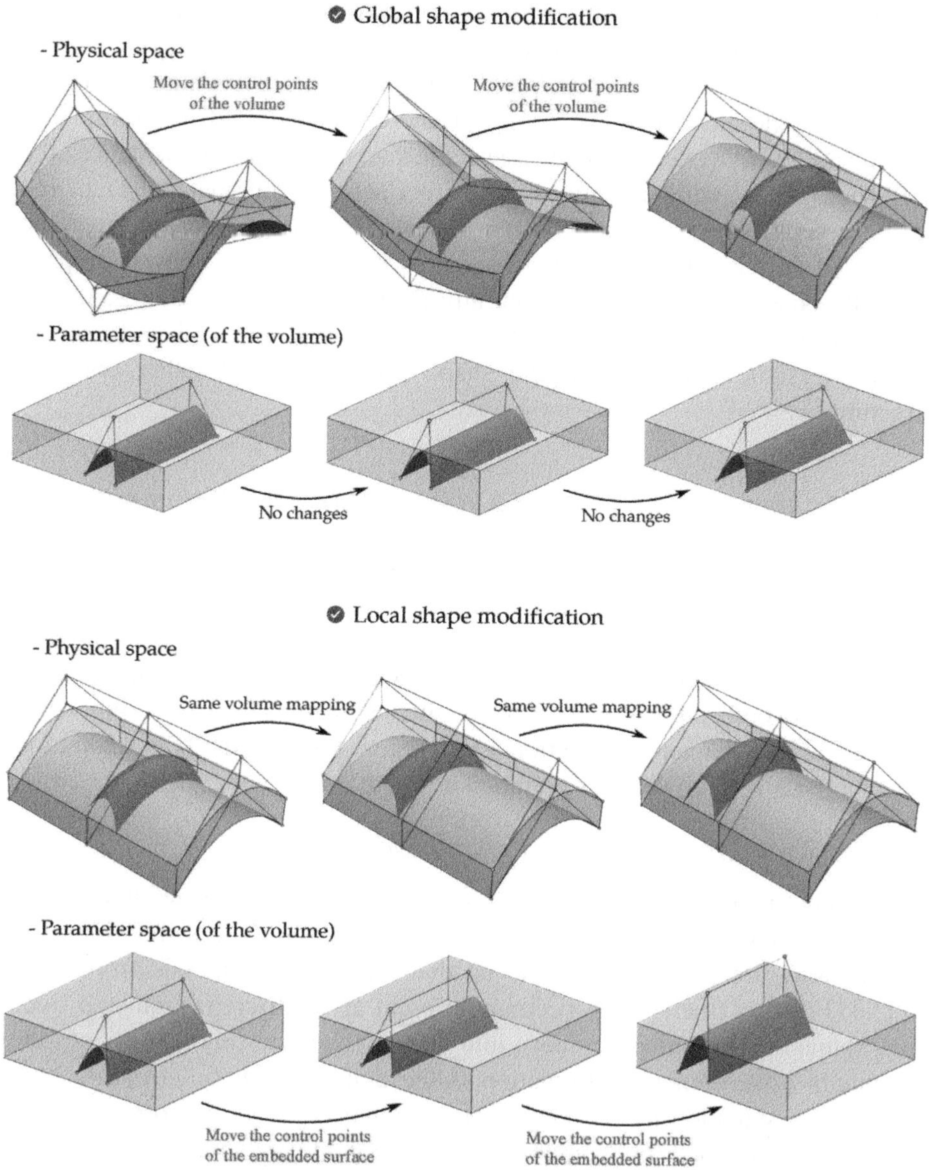

Figure 4.7. *Using spine compositions offers two possibilities for imposing shape deformations. By modifying the volume, we can act on the global shape (FFD-like approach). By acting on the embedded entities, we can modify specific parts, such as, for instance, the stiffeners. The bonding conditions at the interfaces are automatically satisfied. For a color version of this figure, see www.iste.co.uk/bouclier/IGA.zip*

Indeed, using spline compositions offers two possibilities for imposing shape deformations. On one hand, by modifying the volume (through the location of control points P_i), we can act on the global shape. This is, ultimately, an FFD process. On the other hand, by acting on the embedded entities (i.e. moving control points $\bar{P}_j$), we can modify specific parts, such as, for instance, the stiffeners. The bonding conditions at the interfaces are automatically satisfied (from a geometrical point of view). The unidirectional translation of the embedded surfaces will slide the associated surfaces in the physical space along the master surfaces while naturally respecting their curvatures, if any. This point will be further accounted for in section 4.6, which takes advantage of this versatile approach for the shape optimization of rather complex structures.

4.4.2. *Analysis: an embedded Kirchhoff–Love shell element*

Above, we have introduced the use of spline compositions for geometric requirements. In order to carry out the shape optimization process in Figure 4.1, we also need to perform the structural (along with sensitivity) analysis for these surfaces obtained by spline compositions. To this end, we now build an embedded Kirchhoff–Love shell element. The basic components presented in Chapter 3 (section 3.3.1) for the standard isogeometric Kirchhoff–Love shell serve as prerequisites here.

4.4.2.1. *Mid-surface defined by spline composition*

The continuum formulation of the embedded Kirchhoff–Love is formally identical to the standard Kirchhoff–Love formulation. The same kinematic assumptions are postulated. It means that all of the equations from [3.1] to [3.17] remain true. Things become different at the discretization step since the spline composition comes into play. The starting point of the Kirchhoff–Love element formulation is the definition of the covariant base vectors (see equation [3.3]). Returning to equation [4.31], we observed that the surface obtained by composition was parameterized with curvilinear coordinates θ_1 and θ_2. As a result, the definition of the covariant basis vectors is straightforward and is given by the chain rule:

$$A^h_\alpha = \frac{\partial S^h}{\partial \theta_\alpha} = \sum_{k=1,3} \left[\bar{S}^h_{,\theta_\alpha} \cdot e_k\right] V^h_{,\xi_k}\left(\bar{S}^h\right) \quad \alpha = 1,2\,, \qquad [4.32]$$

where the subscript $(\cdot)_{,\xi} = \partial(\cdot)/\partial\xi$ indicates the partial derivative with respect to variable ξ and e_k are the unit basis vectors associated with the volume coordinate system in the parameter space (i.e. corresponding to direction ξ_k; see Figure 4.5 (left)). Then, given equation [3.12], the derivatives of the covariant basis vectors versus the curvilinear coordinates are also required. Starting from [4.32], we thus need to apply the chain rule one more time, which yields:

$$A^h_{\alpha,\theta_\beta} = \sum_{k=1,3} \left[\bar{S}^h_{,\theta_\alpha\theta_\beta} \cdot e_k\right] V^h_{,\xi_k}\left(\bar{S}^h\right) + \sum_{k=1,3}\sum_{l=1,3} \left[\bar{S}^h_{,\theta_\alpha} \cdot e_k\right]\left[\bar{S}^h_{,\theta_\beta} \cdot e_l\right] V^h_{,\xi_k\xi_l}\left(\bar{S}^h\right). \quad [4.33]$$

Regarding the geometric quantities, equations [4.32] and [4.33] constitute the only changes that need to be performed from the standard isogeometric Kirchhoff–Love shell formulation, in order to achieve the new embedded Kirchhoff–Love shell formation. As a result, we actually have to add very little modification to an existing code including the standard Kirchhoff–Love isogeometric elements, so that it offers the embedded version.

NOTE.– It may be noted from equation [4.33] that we need a smooth C^1 surface for the Kirchhoff–Love shell formulation (since second-order derivatives of the surface are involved). This requirement can be satisfied here with two spline mappings of higher regularity. Indeed, if both the embedded surface $\bar{S}^h$ and the volume mapping V^h are C^1, then the spline composition S^h is *de facto* C^1.

4.4.2.2. *Approximation space for the displacement field*

In addition to the discretization of the geometry, we need to clarify how the displacement field is discretized to get the embedded Kirchhoff–Love shell element. In fact, the isoparametric concept may not be applicable in the case of a spline composition. We basically have two choices in the present situation (once again, we make sure that the displacement field is at least C^1 to be consistent with the Kirchhoff–Love formulation; see [3.12]):

1) based on the trivariate spline discretization of the mapping [4.29]:

$$u^h(\theta_1, \theta_2) = \sum_i R_i\left(\bar{S}^h(\theta_1, \theta_2)\right) u_i; \quad [4.34]$$

2) based on the bivariate spline discretization of the embedded surface [4.30]:

$$u^h(\theta_1, \theta_2) = \sum_j \bar{R}_j(\theta_1, \theta_2)\, \bar{u}_j. \qquad [4.35]$$

The first choice provides an unfitted-like approach where the deformation of the shell is prescribed by the surrounding volume (see, for example, Rank *et al.* (2012), Legrain (2013), Schillinger and Ruess (2015) and Chapter 2). The second choice is closer to the standard Kirchhoff–Love isogeometric shell. The mapping only plays a role from the geometric point of view, since it modifies the shape of the embedded surface, but it is not involved in the functional space. We consider the second approach for the intended applications in this work, given its numerical and implementation simplicity over the first alternative (no need for specific quadrature rules, specific treatments to counterbalance the ill-conditioning of the system, etc.). Consequently, the difference, with respect to the standard shell formulation, essentially lies in the additional trivariate mapping step that modifies the shape of the undeformed configuration. In the end, introducing approximation u^h from [4.35] into [3.11] and [3.12] and applying the equilibrium of the shell leads to the typical system [3.18], where the trivariate spline mapping is introduced into [3.19].

Since the final geometric model obtained by spline composition is also directly used in the analysis step through this embedded Kirchhoff–Love shell element, there exists an exact link between the shape parameters and the resulting analysis models (in other words, no approximations are involved during the shape update and transformation into the analysis model). As a result, the proposed strategy is totally adapted to the analytical derivation of the sensitivities, which is one of the crucial weapons of IGA that make it highly relevant for shape optimization (see the discussion in section 4.2.4). In this respect, our approach is robust and generic in contrast to the current FE-based techniques (Jutte *et al.* 2014; Locatelli *et al.* 2014; Dubois *et al.* 2018), where several approximations usually separate the design and analysis models. For details regarding the computation of the analytical sensitivities associated with the embedded approach, we refer the reader to Hirschler (2019) (Chapter 5).

NOTE.– Let us emphasize again that the other crucial ingredient from the analysis point of view of our methodology is the use of the DD solver from Chapter 3, which is adapted to the coupling of non-conforming isogeometric shell patches. In the present situation of the embedded Kirchhoff–Love shell formulation, the application of the DD strategy only requires some precautions regarding the implementation (to take into account the additional trivariate mapping to define the interface) are required. The particular point that has made it possible is the pure algebraic approach followed to define the local null spaces and pseudo-inverses in the DD solver (see section 3.4.3)). Indeed, with the embedded shell formulation being non-isoparametric, it is impossible to build, in general, the kernel of the stiffness operators of the embedded shells *a priori*. We refer the reader to Hirschler *et al.* (2019c) and Hirschler (2019) (Chapter 4) for more information on this point.

4.4.3. *Two preliminary examples to illustrate the design capabilities*

Before applying the proposed strategy for complex shape optimization (see section 4.6), we carry out two simple examples that enable us to clearly understand the benefits of the approach in terms of design capabilities. More precisely, we first solve a shell optimization problem, where a reference analytical solution exists, with the two shape deformation possibilities offered by the embedded strategy. This enables us to validate the two variants of our methodology. Then, we repeat a famous shell shape optimization problem from literature, but with a multipatch structure to account for the potential of our approach to handle complex multipatch models.

4.4.3.1. *FFD versus embedded shape updates*

4.4.3.1.1. Arch under constant vertical load

Let us take the example of the arch simply supported at its two ends and subjected to constant vertical load, as depicted in Figure 4.8. This test case was built in Kiendl *et al.* (2014) so as to provide a shell shape optimization example with an exact solution, and can thus serve as a general benchmark for shape optimization methods. The optimization problem is an unconstrained problem, and the objective is to minimize compliance. Under the consideration of a small thickness t, it can be found analytically that the bending stiffness vanishes when the arch represents a parabola with a specific length-to-height ratio $h/L = 0.54779$. Since a shell is much more rigid when subjected to membrane than to bending, this configuration constitutes the optimal shape. We can solve this arch optimization problem numerically by modeling and computing the arch with standard isogeometric Kirchhoff–Love shell elements. The design variations of the shell are imposed in the vertical direction through design variables, which move the inner control points of the beam, as shown in Figure 4.8. Ensuring to start with a random configuration that is not strictly planar (to avoid being trapped in the maximum of the objective function), we have to find, with the parameters indicated in Figure 4.8, a parabola with a height $h_{\text{opt}} = 5.4779$ m as the optimal solution.

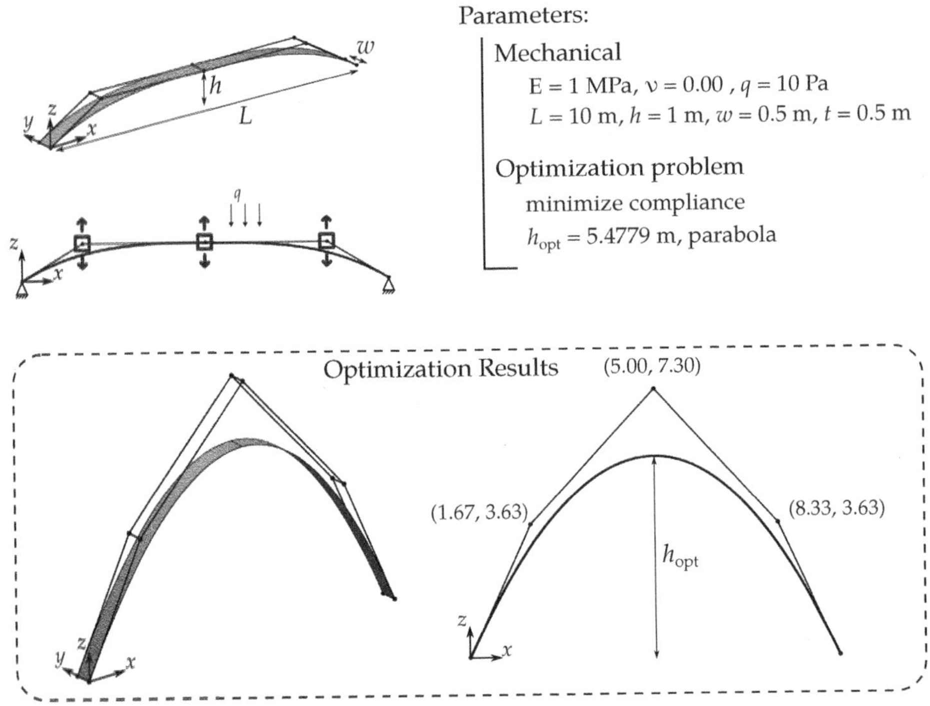

Figure 4.8. *Settings for the shape optimization of an arch under constant vertical load. The optimal result is a parabola with a specific height-to-length ratio (see Kiendl et al. (2014)). The problem can be solved while using at least one quadratic B-spline element for the design model, and classical Kirchhoff–Love shell elements for the structural analysis. In order to do so, a set of design variables (three in the illustration corresponding to the use of two quadratic elements for the design model) move couples of control points in the vertical direction*

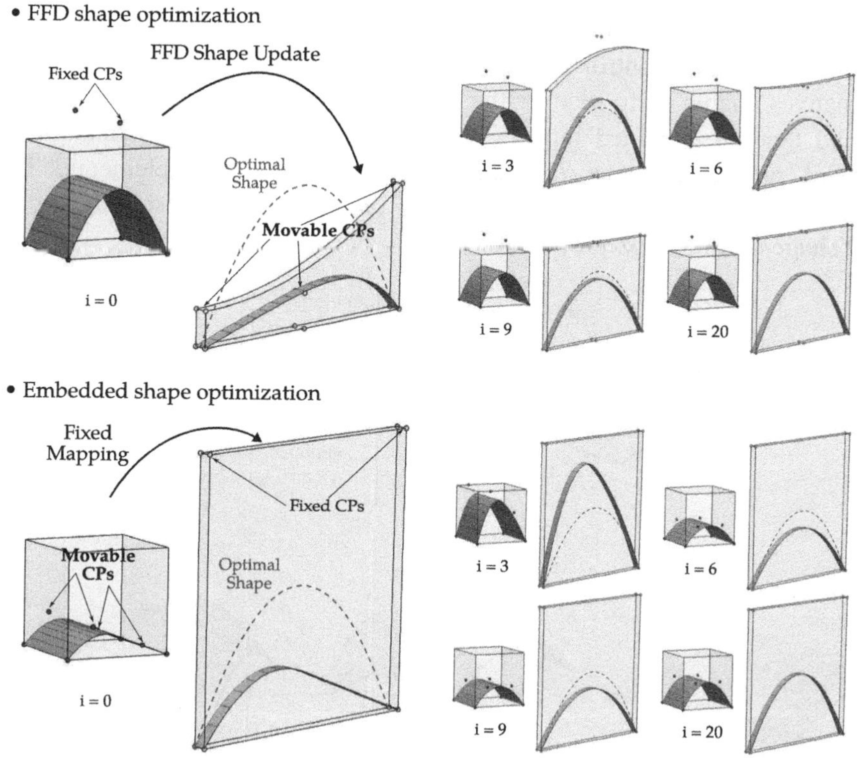

Figure 4.9. *Resolution of the arch optimization problem with embedded Kirchhoff–Love shell elements (top) in the case of FFD-based shape updates, and (bottom) in the case of embedded shape updates. The dots represent the control points of the design models (of the surface and the volume), while the lines on the arch correspond to the element boundaries of the analysis model. For a color version of this figure, see www.iste.co.uk/bouclier/IGA.zip*

4.4.3.1.2. Resolution with the two variants of the embedded strategy

Now, let us try to solve the arch optimization problem with the developed embedded strategy. The difference lies in the imposition of the shape variations. Reminded of the potential of the approach for shape optimization (illustrated in Figure 4.7), this can actually be done in two ways. Figure 4.9 presents the results of the resolution of the arch optimization problems with the two associated variants. First, the design variables can be related to the control points of the volume mapping (see Figure 4.9 (top)). We call this approach the *FFD shape update* because it recalls FFD techniques. Second, the design variables can be linked to the control points of the embedded surface in order to modify the shape of the shell (see Figure 4.9 (bottom)).

This second approach is referred to as the *embedded shape update* since these are the embedded control points that move. As it can be seen, the two alternatives enable us to recover the optimal solution. Furthermore, it has to be said that the obtained number of iterations is equivalent to that of the standard procedure using classical Kirchhoff–Love shell elements. This validates the two possibilities offered by our approach for solving typical shell shape optimization problems.

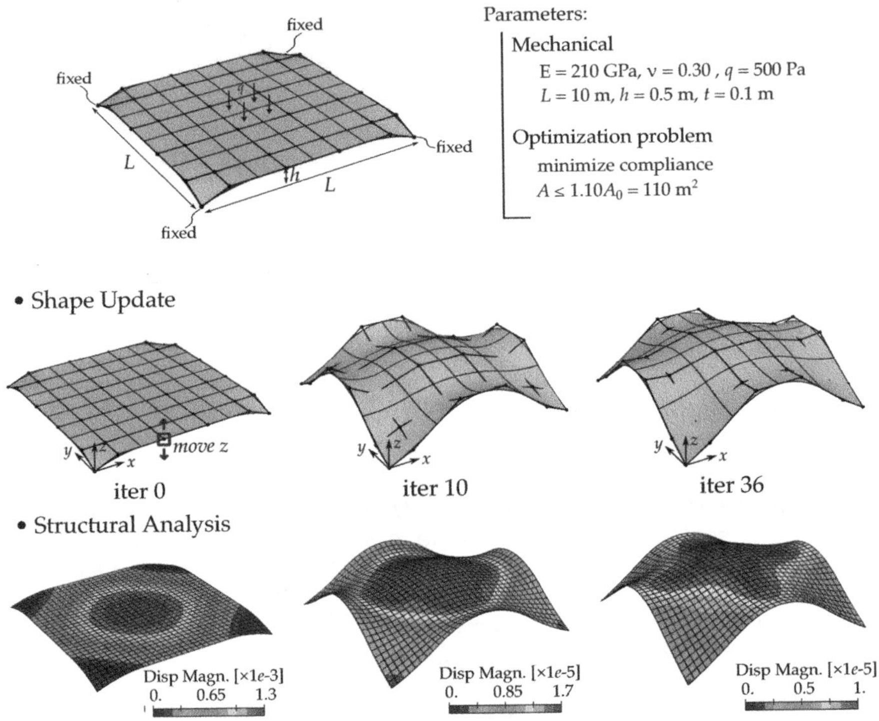

Figure 4.10. *Settings and optimization results for the initial square roof problem. The design model is defined by a single quadratic B-spline patch, which discretizes the whole structure in 4-by-4 elements. The surface is parameterized with 32 design variables; the four control points located at the corners are kept fixed. A refinement level of 3 is applied in both parametric directions, resulting in an analysis model with 1024 quadratic elements. The area is constrained to be lower than 110% of the initial area. For a color version of this figure, see www.iste.co.uk/bouclier/IGA.zip*

Obviously, other parameterizations could have been considered to solve the problem. For the FFD shape update, we chose to define the embedded surface as a parabola (a single B-spline quadratic element for the design model) with an arbitrary length-to-height ratio. As a result, a linear mapping, used to obtain the optimal "volume" from the initial one, was expected to rectify the ratio of the parabola. This is what we obtained while considering quadratic functions in the longitudinal direction (and linear functions in the other directions) for the design model of the volume. Note that initializations other than parabolas could have been considered and would have also led to the correct optimal design. In the case of embedded shape updates, the volume is fixed. We thus made sure to make it sufficiently large to ensure that the design space would contain the optimal solution. The design model for the embedded surface was composed of two quadratic elements, while only one linear element was sufficient for the design model of the volume in this case. For the structural analysis, we discretized the arch with 16 quadratic B-spline elements.

4.4.3.2. *Optimizing the global shape of a multipatch structure*

4.4.3.2.1. Initial square roof problem

As a second example to highlight the design capabilities of the proposed embedded approach, we investigate a multipatch version of the square roof problem described in Figure 4.10 (top). Let us first present the initial, one-patch, test case before entering into the details of the multipatch variant (see Figure 4.13). The example is based on the work of Kegl and Brank (2006), and similar studies can be found, for example, in Bletzinger *et al.* (2005) and Ikeya *et al.* (2016). The roof is fixed on its four corners and is subjected to a uniform snow load (see Figure 4.10 for the different values of the model parameters). Starting with a random configuration, which is not rigorously planar, the objective is to optimize the shape of the structure by minimizing the compliance under a maximal area constraint. Using the standard isogeometric shape optimization framework along with classical Kirchhoff–Love shell elements, we are able to obtain similar results to those in Kegl and Brank (2006), as shown in Figure 4.10. The design variations of the shell are imposed in the vertical direction through design variables that move each of the control points, except those at the four corners. The compliance is drastically reduced from a (planar) plate with a factor of $1.44 \cdot 10^{-3}$. For more information on the obtained solution, the optimal positions of the control points are given in Figure 4.11. The final shape has four symmetric planes defined. This is expected as the problem presents these symmetries. Therefore, some design variables have identical optimal values. The fact that we obtain the symmetries without setting groups of design variables is a meaningful indication to validate the result.

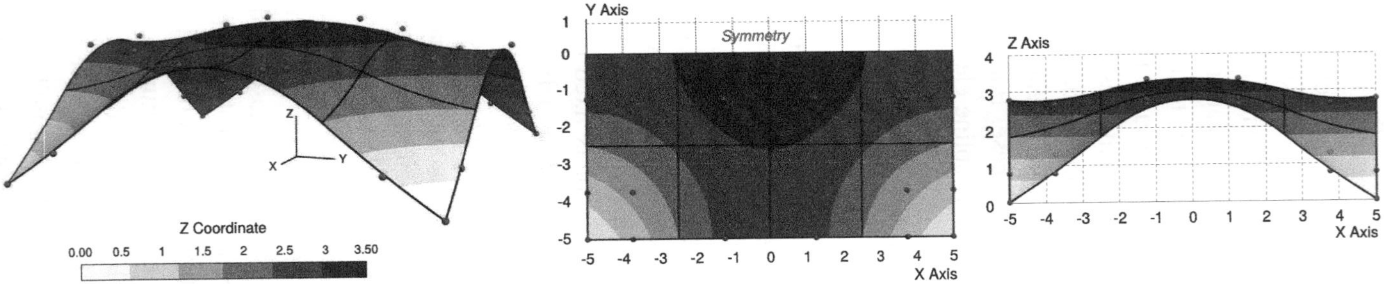

Figure 4.11. *Optimal shape of a square roof subjected to an uniform snow load: (left) 3D view, (middle) top view and (right) side view of the final structure. The red points represent the control points. For a color version of this figure, see www.iste.co.uk/bouclier/IGA.zip*

4.4.3.2.2. The multipatch version solved with the FFD shape update approach

Our objective now is to demonstrate the capability of our embedded approach to optimize the global shape of a multipatch structure. To this end, we repeat the square roof test case, but with a multipatch model. As an example, take a square plate made of two non-matching patches, as described in Figure 4.12 (left). It is straightforward that the control points of each patch cannot freely move at the interface (see Figure 4.12 (right)), but on the contrary, some constraints need to be imposed such that a certain level of continuity is preserved at the interface during the shape updates. This question is not new and solutions have already been studied, such as the weak enforcement of geometric continuity (G^1-continuity) through the introduction of variational equality constraints (Schmidt 2013). Here, we apply the global optimization approach (or FFD shape update) depicted in Figure 4.7 (top). It has the strong advantage of explicitly taking into account the adequate geometric constraints at the different multipatch interfaces of a complex design model.

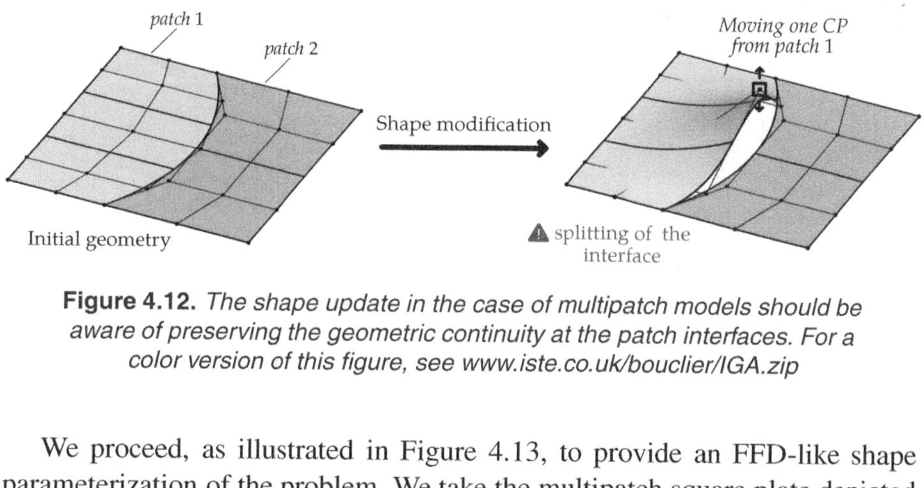

Figure 4.12. *The shape update in the case of multipatch models should be aware of preserving the geometric continuity at the patch interfaces. For a color version of this figure, see www.iste.co.uk/bouclier/IGA.zip*

We proceed, as illustrated in Figure 4.13, to provide an FFD-like shape parameterization of the problem. We take the multipatch square plate depicted in Figure 4.12 (left) and perform mesh refinement to arrive at an analysis model composed of 33-by-16 (patch 1) and 32-by-16 (patch 2) cubic B-spline elements. Their common interface describes a parabola of height $h = 2$ m and length $L = 10$ m. Obviously, the mechanical coupling is applied using the Mortar approach, as presented in section 3.3. For the design model, a box is built that surrounds the plate. It consists of a B-spline volume (made of 8-by-8-by-1 quadratic elements) that represents the identity transformation, i.e. its parameter and physical spaces are identical. Hence, the plate can now be seen as an embedded surface in our methodology (see equation [4.31]). We impose shape modifications to the plate by modifying the volume mapping (equation [4.29]). Taking the volume as a quadratic (and thus C^1) B-spline ensures that the geometric continuity at the non-matching multipatch interface is preserved. More precisely, we only move the control points from the upper face of the box (expect those in common with the other faces) vertically. It leads to 64 design variables. Of course, the definition of the volume is not unique and other choices could have been carried out. The presented configuration has the advantage that the four corners of the plate are located on the faces of the box. As a result, if we do not want to modify the location of these corners, we can simply keep these faces untouched. We make this choice here.

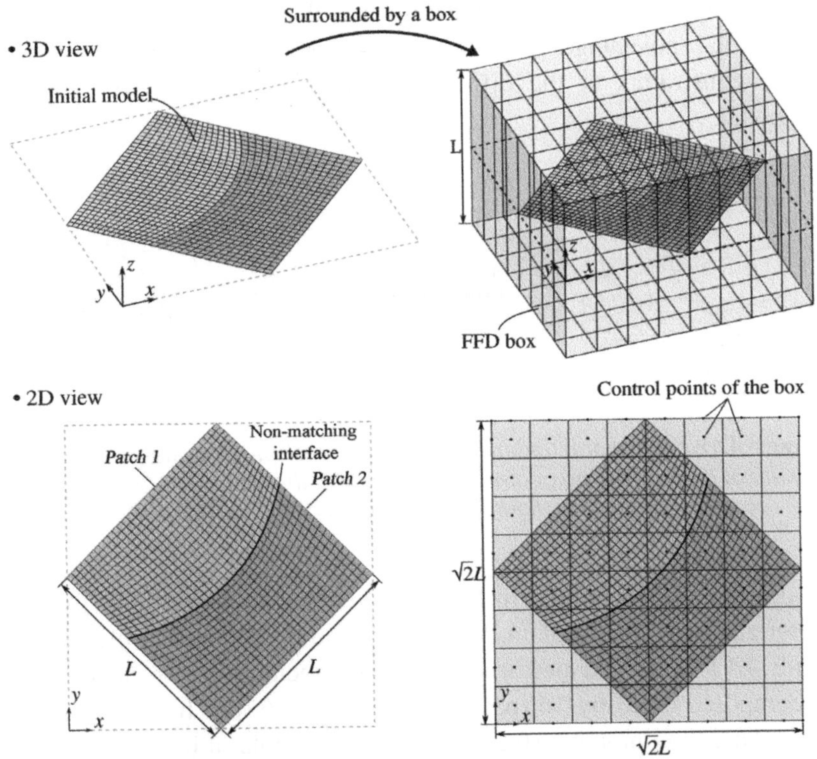

Figure 4.13. *Setting up an FFD-like shape parameterization in order to impose shape modifications that preserve the geometric continuity between the non-matching patches. For a color version of this figure, see www.iste.co.uk/bouclier/IGA.zip*

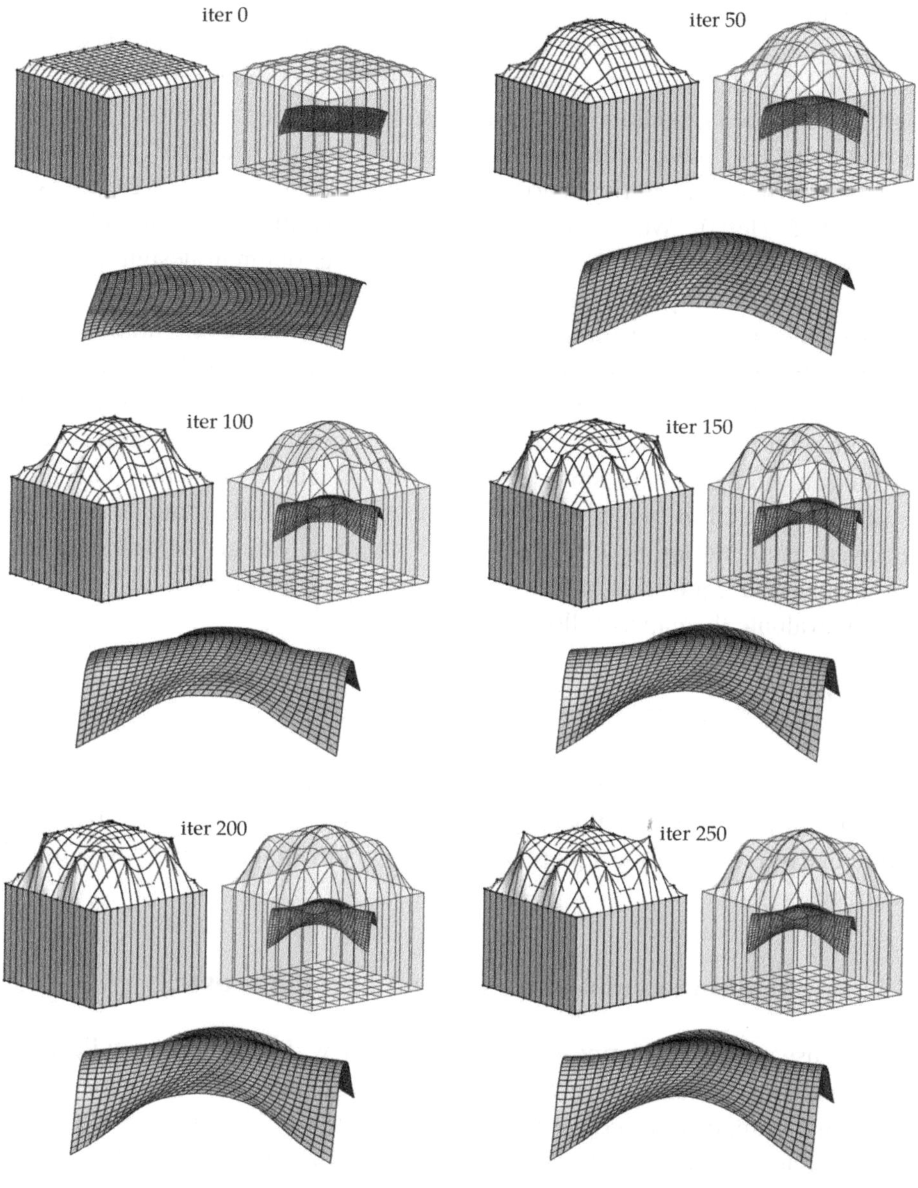

Figure 4.14. *Optimization history of the FFD shape update approach applied to the shape optimization problem of the square roof with a multipatch model. For a color version of this figure, see www.iste.co.uk/bouclier/IGA.zip*

Figure 4.14 shows the numerical results obtained for the shape optimization problem of the multipatch square roof with our FFD shape update approach. We recover a similar shape to the one obtained with a monopatch model and the standard Kirchhoff–Love shell formulation. To highlight this point, additional numerical tests are carried out in Figure 4.15. In particular, we present the results obtained with the standard IGA-based approach, where the optimization model counts 96 design variables (see Figure 4.15 (left)). We consider this optimal geometry as the reference solution. The differences between this reference optimal design and the non-matching multipatch result (see Figure 4.15 (right)) are almost undetectable. Looking at the displacement fields of both configurations also confirms the correctness of the multipatch resolution. Finally, in order to study the influence of having a multipatch model, we solve the optimization problem with the FFD shape update approach applied to a monopatch model. We take a simple square surface with 32-by-32 cubic elements. The final result is presented in Figure 4.15 (middle). Again, a similar optimal shape is achieved. By denoting by $\mathbf{s}^{\text{multi}}$ and $\mathbf{s}^{\text{mono}}$ the final design variables for the multipatch and equivalent monopatch optimization resolution, respectively, we can evaluate the gap as follows:

$$\frac{||\mathbf{s}^{\text{multi}} - \mathbf{s}^{\text{mono}}||}{||\mathbf{s}^{\text{multi}} - \mathbf{s}^{\text{mono}}||} = 0.137\%,$$

which definitely accounts for the well-handled multipatch models. Note that a slight asymmetry exists for the multipatch case, which originates from the initial discretization of the square plate that is not symmetric due to the curved interface.

4.5. Application to solid structures and first interests

The use of spline compositions has a much wider range of applications than modeling stiffened aerostructures. Indeed, geometric modeling through functional composition has recently gained interest in several fields, and it is not restricted to thin structures. As an example, models of complex micro-structures through the use of compositions of trivariates (i.e. of volumes) can be found in Elber (2017) and Antolin *et al.* (2019). In this section, we discuss the generalization of the methodology proposed in section 4.4 for the proper and efficient shape optimization of such complex full 3D solid structures.

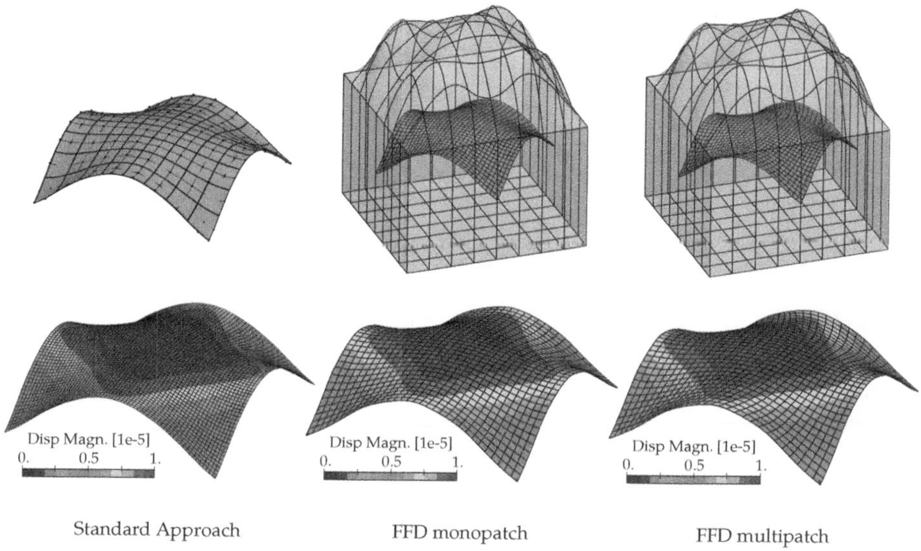

Figure 4.15. *Comparison of the results obtained for the shape optimization of the square roof. Our FFD shape update approach applied to the multipatch model gives very similar results to both monopatch resolutions (standard and FFD shape update). For a color version of this figure, see www.iste.co.uk/bouclier/IGA.zip*

4.5.1. *Simple extension of the method*

An example of the composition of two volumes is depicted in Figure 4.16. Everything is rather similar to the modeling approach introduced above for stiffened aerostructures (section 4.4.1.2). The only prerequisite for being able to perform the composition is that the image of the first mapping is contained in the parameter space of the second mapping, as shown in Figure 4.16. As a result, we could also potentially embed curves into surfaces or into volumes. Such cases can actually be found in Elber (2017), for example. Focusing here on the composition of volumes, we start by defining the following two spline trivariate mappings:

$$V^{1h}(\theta_1, \theta_2, \theta_3) = \sum_{i=1}^{n_{3D}} R_i^1(\theta_1, \theta_2, \theta_3) P_i^1, \quad \forall(\theta_1, \theta_2, \theta_3) \in \bar{\Omega}^1, \qquad [4.36]$$

$$V^{2h}(\xi_1, \xi_2, \xi_3) = \sum_{j=1}^{m_{3D}} R_j^2(\xi_1, \xi_2, \xi_3) P_j^2, \quad \forall(\xi_1, \xi_2, \xi_3) \in \bar{\Omega}^2, \qquad [4.37]$$

and we compose them in order to get the expression of the complete mapping, as done previously for embedded surfaces (equation [4.31]). The composed volume reads as:

$$V^{(1\circ 2)h}(\theta_1, \theta_2, \theta_3) = \left(V^{2h} \circ V^{1h}\right)(\theta_1, \theta_2, \theta_3), \quad \forall(\theta_1, \theta_2, \theta_3) \in \bar{\Omega}^1. \qquad [4.38]$$

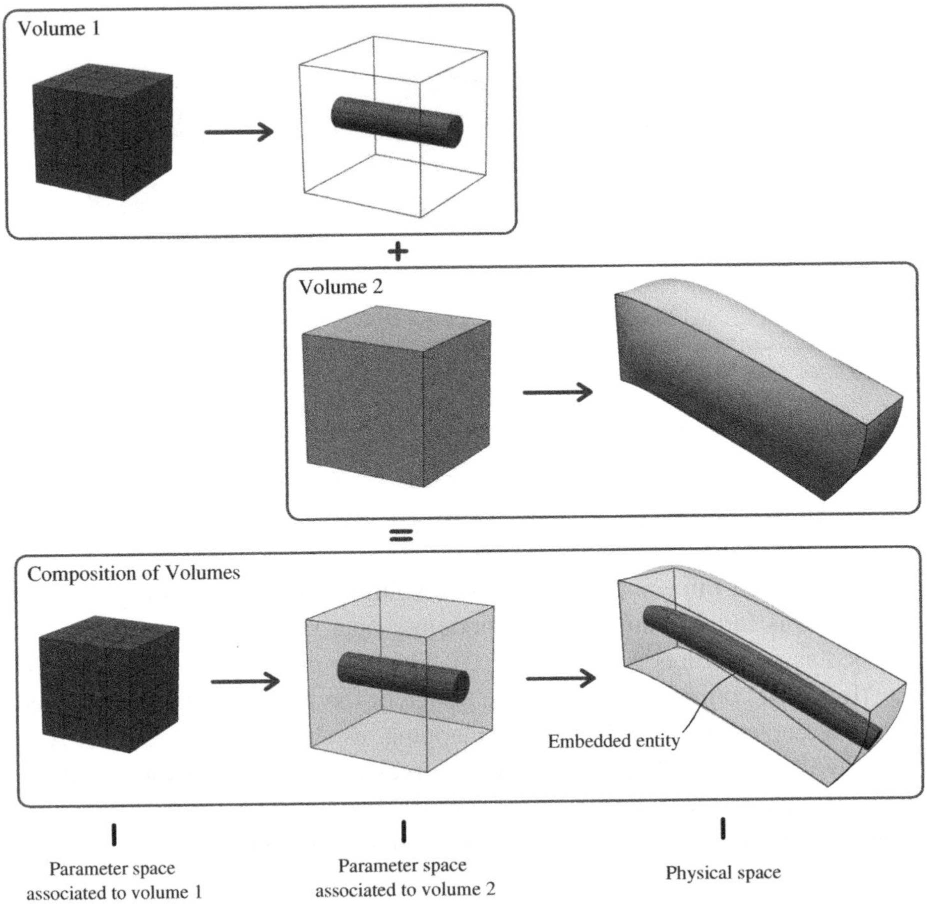

Figure 4.16. *Composition of trivariates for proper complex solid geometries*

This volume takes the same parameters as the first volume and maps them into the image space of the second volume. In the specific case of a simple second volume mapping (such as a Bézier mapping, i.e. roughly speaking, a one-element B-spline mapping), the composition can be rewritten using a single spline mapping, but of a (very) high degree (Elber 2017). In more general situations, for example, when the first volume crosses the knot lines associated with the second volume, the resulting composed mapping is no longer a spline tensor product (van Sosin and Elber 2018).

Regarding the analysis, we must pay attention to the definition of the (discrete) solution space. Indeed, applying the isoparametric principle is not trackable in practice. Given the previous discussion, it could actually be possible in the very specific case of a composition that results in a single spline mapping, but it would lead to a computationally demanding model (in particular, with many integration points), since the spline mapping would be of a very high polynomial degree. Consequently, the same discussion as for the embedded shell formulation (remind section 4.4.2.2) has to be carried out. We can opt for an unfitted-like approach by using the basis functions associated with the surrounding volume (or possibly a refined version of this space):

$$u^{(1\circ 2)h}(\theta_1, \theta_2, \theta_3) = \sum_j R_j^2\big(V^{1h}(\theta_1, \theta_2, \theta_3)\big)u_j. \qquad [4.39]$$

Alternatively, we can recycle the discrete space associated with the first volume (with an additionally refinement step, if needed). In this case, the solution is of the following form:

$$u^{(1\circ 2)h}(\theta_1, \theta_2, \theta_3) = \sum_i R_i^1(\theta_1, \theta_2, \theta_3)\big)u_i. \qquad [4.40]$$

As for the embedded shell formulation, in the following we use in the second variant [4.40], which is much simpler from an implementation point of view. The second mapping, which is the FFD box mapping, only plays a role in the evaluation of the integrals to build the stiffness matrix. Once the solution space is defined, the analysis procedure follows the considered finite element workflow. Consequently, these types of formulation could easily be integrated into any code that can handle non-isoparametric elements.

4.5.2. *A test case in 2D*

Let us imagine the following simple toy case, in order to highlight the interest of using functional composition for shape optimization. We have, in hand, a non-matching multipatch model of the cantilever beam, such as those investigated in section 3.5.1.1. For example, let us consider that the goal is to perform design optimization with the model denoted `1C` in Figure 3.8. Due to the non-matching interfaces, we will face some difficulty in defining a shape parameterization that preserves the inter-patch geometric continuity. We would need to define a relationship between the control points of adjacent patches. While this remains possible, we present an alternative approach that highly simplifies the enforcement of those constraints.

The idea consists of introducing a new global mapping that embeds the initial multipatch model, as described in Figure 4.17. We actually apply the same strategy that has been used previously for the square roof problem (see Figures 4.13–4.15). The FFD box defines an identity map that surrounds the cantilever beam. There are no particular rules to build such a mapping. Here, we decide to define the box such that it takes the same shape as the underlying geometry, namely a rectangle with the exact same dimensions as the beam: $L = 20\,\text{m}$, $h = 5\,\text{m}$ (i.e. as given in Figure 3.8). This new FFD mapping only acts on the geometry. In other words, all of the mechanical parameters (material, boundary conditions, etc.) are unchanged and remain defined on the initial model. We again refer the interested reader to Figure 3.8 for further information on the structural analysis. The only change concerns the prescribed loading: instead of the uniformly distributed load, the beam is this time subjected to a concentrated load $F = 50\,\text{kN}$ at its bottom-right corner.

Finally, setting up the model is rather straightforward. The only modeling choice that remains is the refinement level of the FFD box and the definition of the design variables. In the example depicted in Figure 4.17, the degree is increased from linear to quadratic in the main direction of the beam. We also perform knot insertions, such that the final FFD box counts seven elements uniformly distributed, again, along the main direction of the beam. No refinement is done along the beam height. One design variable is prescribed per control point lying on the upper edge (see Figure 4.17). They move these control points along in the vertical direction. These modeling choices lead to a problem with nine design variables. The design goal is to minimize the compliance without increasing the mass of the structure (prescribed as an inequality equation verified by the volume of the structure). We can find identical optimization problems to this cantilever beam problem using a single-patch model, like the one used in, for example, Wall *et al.* (2008), Qian (2010) and Nagy *et al.* (2013).

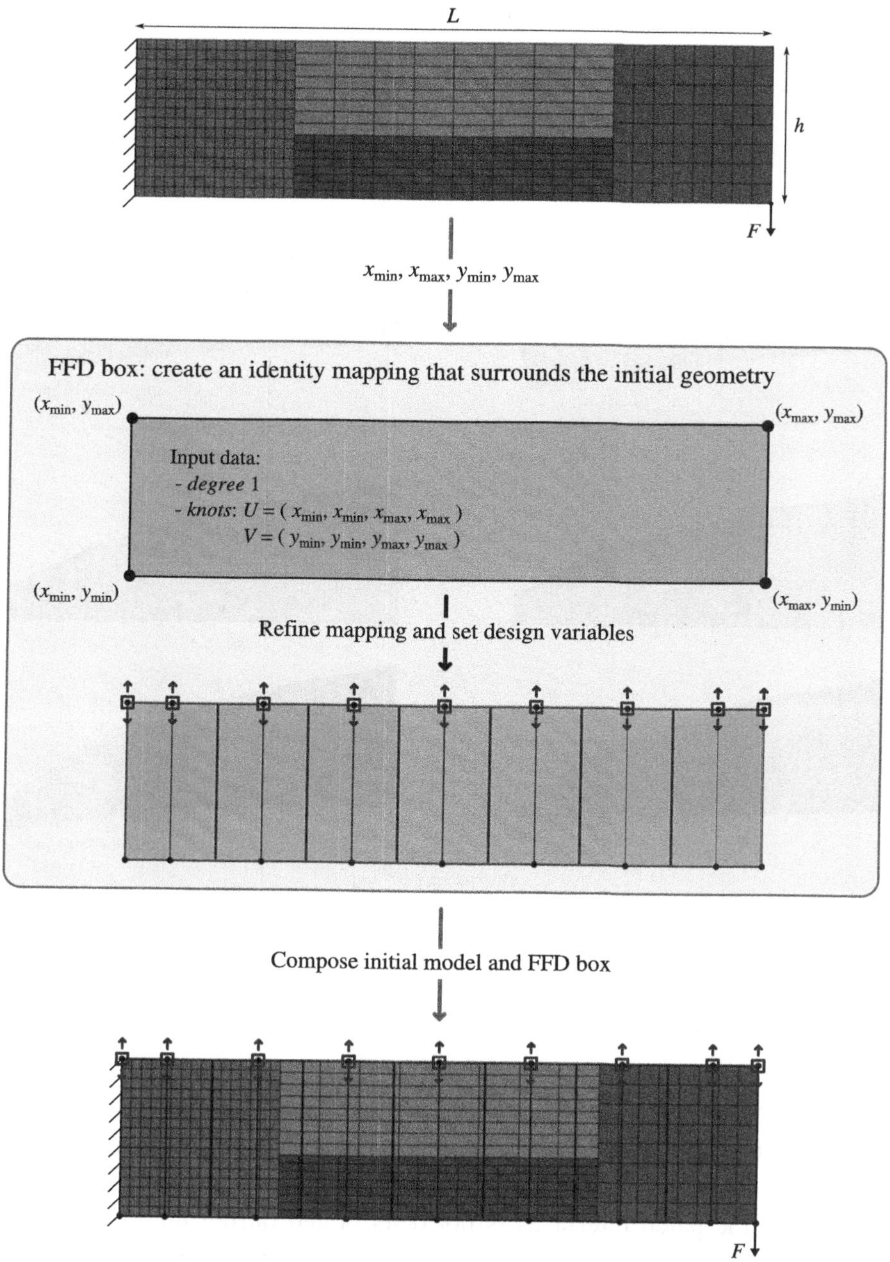

Figure 4.17. *Transforming a non-matching multipatch model into an optimization suitable model by embedding the initial model inside of a free-form deformation box. For a color version of this figure, see www.iste.co.uk/bouclier/IGA.zip*

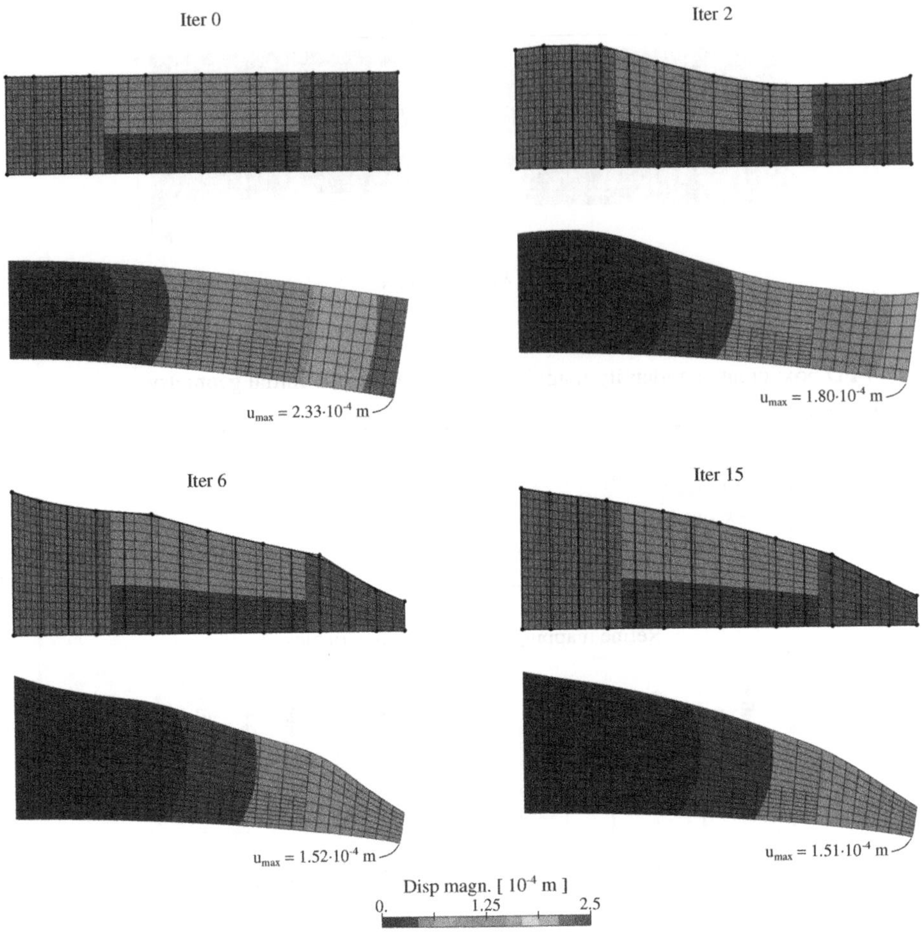

Figure 4.18. *Design optimization of a cantilever beam using the FFD-based approach for general non-conforming multipatch models. For a color version of this figure, see www.iste.co.uk/bouclier/IGA.zip*

The results of the design optimization of the cantilever beam are presented in Figure 4.18. We can observe that the shape updates never introduce gaps at the patch interfaces. Due to the construction, the C^0-continuity of the geometry at the patch interfaces is not deteriorated during the optimization. More interestingly, this is automatically carried out: there is no need to introduce a constraint to enforce this geometric condition. We end up with a beam that has an upper edge with a progressive curvature. This optimal shape reminds us of those obtained by the aforementioned research works where similar optimization problems are conducted, i.e. Wall *et al.* (2008), Qian (2010) and Nagy *et al.* (2013). The number of iterations required by the optimization algorithm to reach convergence is also rather small (a total of 15 iterations). It confirms that this approach based on an FFD strategy behaves well. The optimization procedure does not encounter particular difficulties due to the proposed modeling choices. It leads to identical results that we would have with a standard single model, together with a traditional IGA-based shape optimization framework.

Finally, through this simple example, we show how the use of spline compositions enables us to transform a complex multipatch model with non-conforming interfaces into an optimization-suitable model. From the standard practice in (FE-based) FFD shape optimization (Lyu and Martins 2015; Bletzinger 2017), where the morphing box only acts on the FE nodes (or control points with a simple extend), we quote again that the novelty consists of using real spline compositions, in the sense that the composition is performed over the whole embedded domain. The use of such spline compositions is crucial. Indeed, the standard FFD techniques would fail to impose shape updates that guarantee the inter-patch geometric continuity, while our approach automatically takes those desired constraints into account.

4.6. Advanced numerical optimization examples

We now show that introducing embedded entities into the IGA-based shape optimization framework enables us to tackle challenging optimization problems in a proper and accurate manner. More precisely, we apply the developed approach to three problems of stiffened structures. We start with a stiffened version of the square roof test case to illustrate that the FFD shape update approach is also totally adapted to the specific case of the global shape optimization of stiffened structures. We will see that the global shape modifications of the roof are automatically transferred to the stiffeners. In the next step, a curved wall problem is tackled to show how the design modifications of the embedded surfaces make it possible to optimize the shape and position of stiffeners along curved and complex geometries. In other words, with this second test case we will demonstrate that the embedded shape update strategy offers a smooth way to optimize the local parts of a global shell structure. To conclude, we eventually design an innovative aircraft wing-box structure that proves the applicability of our approach to design futuristic real aeronautical structures.

4.6.1. *Global shell optimization: stiffened roof*

4.6.1.1. *Settings and geometric modeling*

The problem of the stiffened roof is derived from the initial shape optimization problem of a square roof under vertical uniform load (see Figure 4.10 and, for example, Bletzinger *et al.* (2005) and Kegl and Brank (2006)). Here, two stiffeners are added to the initial square plate and the model parameters are taken as indicated in Figure 4.19 (top). By making use of our geometric modeling, we introduce two embedded surfaces to define the stiffeners (see Figure 4.19 (bottom)). The construction consists of two steps. First, the volume mapping is generated by the extrusion of the square plate. We set the degree to one in the direction of extrusion; therefore, the volume shares half of its control points with those associated with the square plate and the additional control points are a copy of the previous ones, with an offset value in the vertical direction. The second step involves embedding surfaces into the parameter space of the created volume. These embedded surfaces, once composed with the volume mapping, give the final surfaces within the physical space that define the stiffeners. Here, the initial embedded surfaces are B-splines of one single element with degree 2-by-1. The coordinates of their control points can be directly interpreted from Figure 4.19, considering the ratio $d_1/d_2 = 0.35$.

In this problem, only global optimization is performed, i.e. the embedded surfaces remain unchanged during the optimization. Therefore, the shape modifications of the stiffeners are only driven by the design updates of the surface representing the roof. For this purpose, we apply the FFD shape update strategy (illustrated in Figure 4.7 (top) and already carried out), for example, for the multipatch square roof in Figure 4.14. The design variables are associated with the control points of the volume mapping. The volume mapping is discretized in 4-by-4-by-1 elements. Degree 3 is taken in directions ξ_1 and ξ_2, while degree 1 is kept in direction ξ_3 (see the red points in Figure 4.19). In the end, the optimization model has 45 design variables that move the control points of the mapping in the vertical direction z. The analysis model is defined such that the roof and both stiffeners are discretized into 1,024 and 256 bi-cubic elements, respectively. As for the original test case, the area of the roof without the stiffeners is constrained to be lower than 110% of the initial area of the square plate.

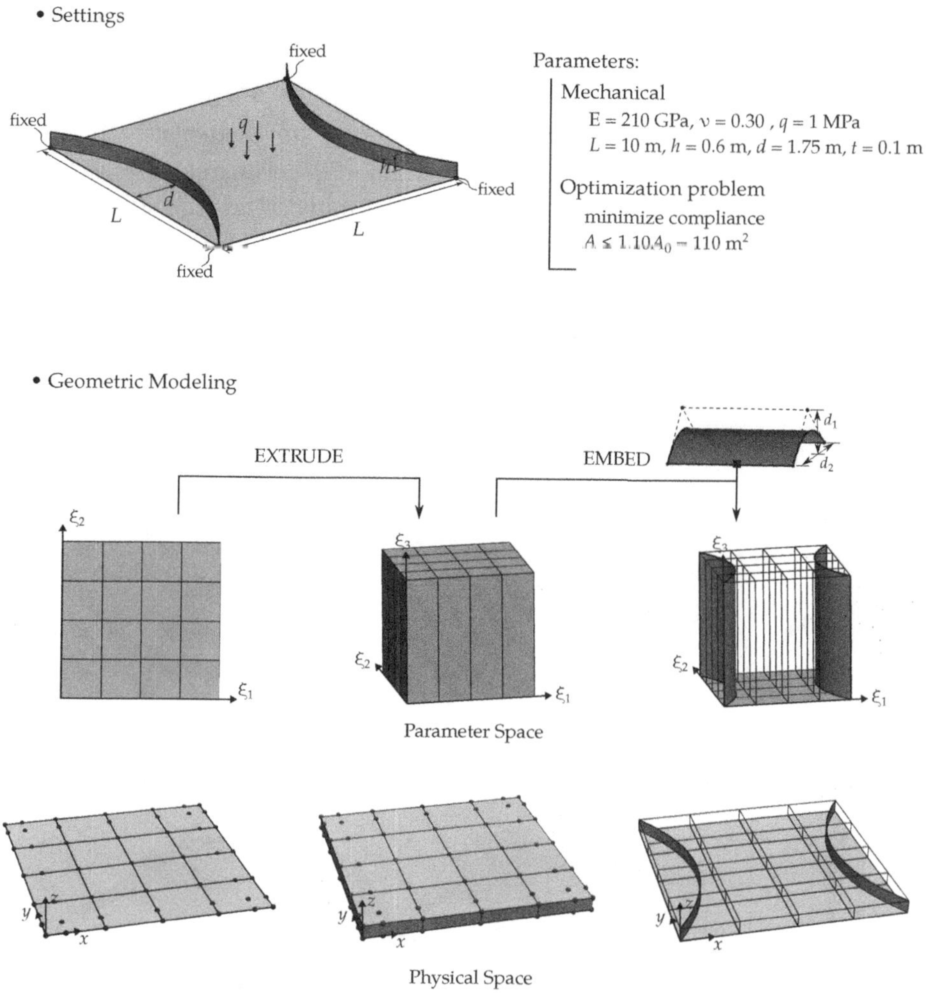

Figure 4.19. *Settings for the stiffened roof problem. The geometric modeling employs the developed embedded approach: a volume is extruded from a square plate, and then, the stiffeners are generated by embedding surfaces into the parameter space of this volume*

4.6.1.2. *Results and discussion*

The shape evolution of the design and analysis models is presented in Figure 4.20 (top). Due to the construction, the stiffeners follow the shape modifications of the roof exactly and naturally during the whole process. The final shape is obtained after 125 iterations of the optimization algorithm when setting the stopping criteria regarding the stagnation of the objective function to $1 \cdot 10^{-8}$. In addition, the evolution of the relative compliance during the resolution of the optimization problem is shown in Figure 4.21(c). The compliance is drastically reduced from the initial stiffened plate: the final relative compliance is equal to $C_{\text{opt}}/C_0 = 2.217 \cdot 10^{-3}$. The final displacement field is presented in Figure 4.21(c). As for the initial ("unstiffened") square roof problem, we recover two planes of symmetry and do not set groups of design variables, which is meaningful in validating the result.

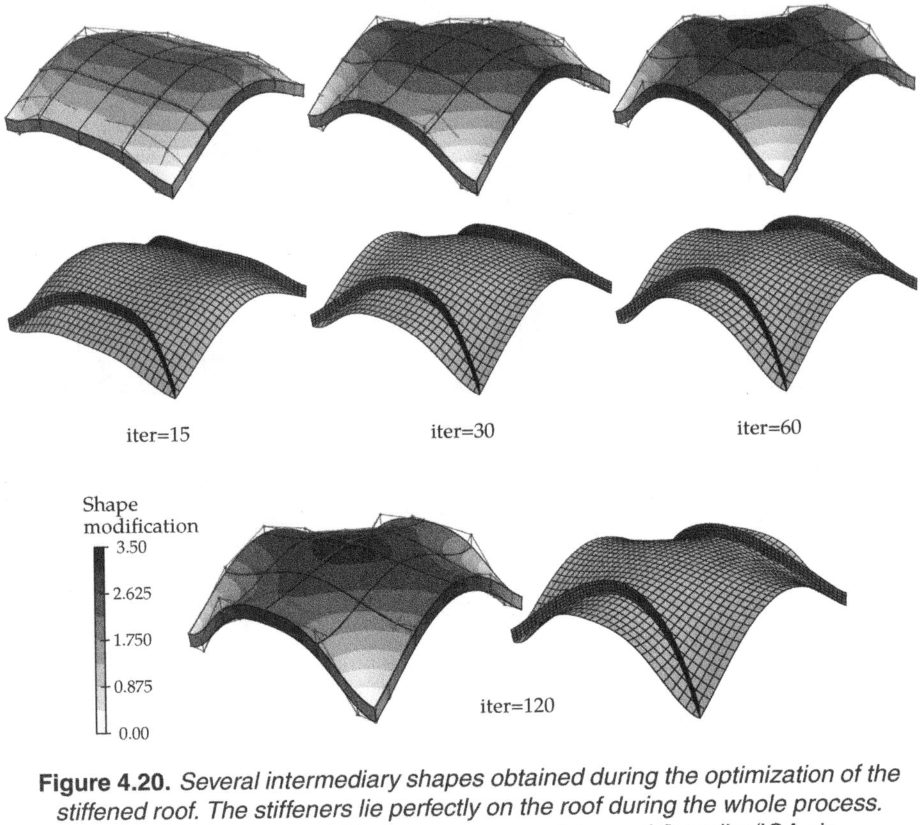

Figure 4.20. *Several intermediary shapes obtained during the optimization of the stiffened roof. The stiffeners lie perfectly on the roof during the whole process. For a color version of this figure, see www.iste.co.uk/bouclier/IGA.zip*

From an engineering point of view, this problem of the stiffened roof highlights an interesting issue. We managed to solve the optimization problem with the substructures. We could simplify the problem by ignoring the stiffeners in the optimization process, assuming that they only have a small influence on the behavior of the global structure. We actually tried to solve the optimization problem without the stiffeners first, and we built the substructure afterwards. We computed the compliance on this new global structure, and we obtained a compliance approximately 15% higher than the one obtained when optimizing the structure incorporating the stiffeners. Given the simplicity of this structure, it highlights that during the optimization, sub-parts such as stiffeners, holes and other geometric details can be essential in designing more complex structures that are better quality.

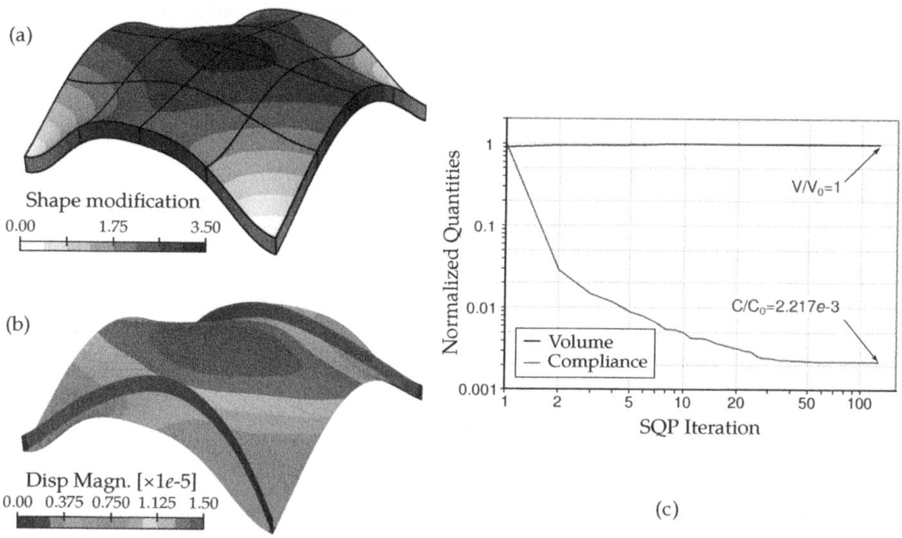

Figure 4.21. *Final design and result for the stiffened roof problem once the algorithm has converged. The optimal design is much stiffer than the initial one. For a color version of this figure, see www.iste.co.uk/bouclier/IGA.zip*

4.6.2. *Local shell optimization: curved wall*

The second problem addresses the optimization of the shape and the position of a complex "cylindrical" stiffener lying on a curved surface that represents a wall. Contrary to the stiffened roof problem, we do not modify the geometry of the wall here, but rather perform a local optimization. For this problem, we only focus on the substructure. We thus apply the embedded shape update strategy illustrated in Figure 4.7 (bottom), i.e. we only move the control points of the embedded surface (that are defined in the parameter space of the volume).

• Settings

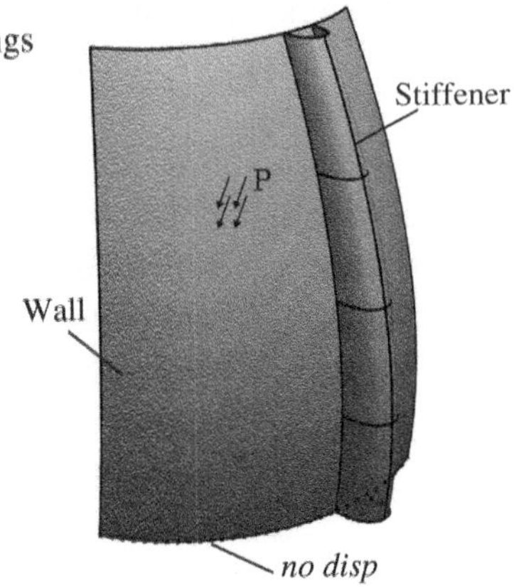

Parameters:

Mechanical

$E = 100$ GPa, $\nu = 0.30$, $P = 0.1$ MPa

$t_w = 0.25$ m, $t_s = 0.1$ m

Optimization problem

minimize compliance

$A \leq A_0$

• Geometric Modeling / Shape Parametrization

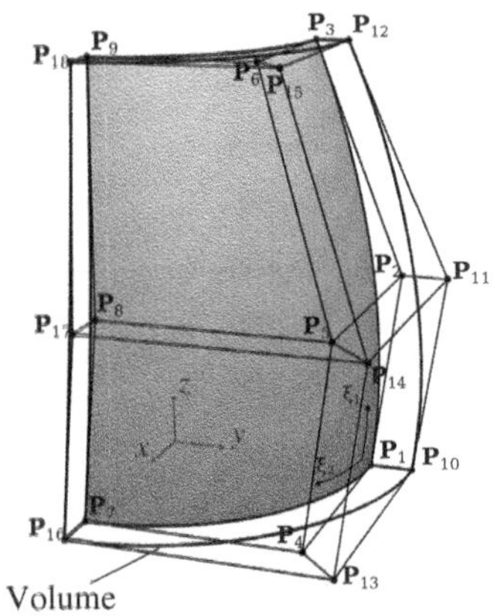

x	y	z	w
0.	5.	0.	1.
1.	6.	5.	1.
0.	3.75	10.	1.
5.	5.	0.	$1./\sqrt{2}$
6.	6.	5.	$1./\sqrt{2}$
3.75	3.75	10.	$1./\sqrt{2}$
5.	0.	0.	1.
6.	1.	5.	1.
3.75	0.	10.	1.
0.	6.	0.	1.
1.	7.	5.	1.
0.	4.5	10.	1.
6.	6.	0.	$1./\sqrt{2}$
7.	7.	5.	$1./\sqrt{2}$
4.5	4.5	10.	$1./\sqrt{2}$
6.	0.	0.	1.
7.	1.	5.	1.
4.5	0.	10.	1.

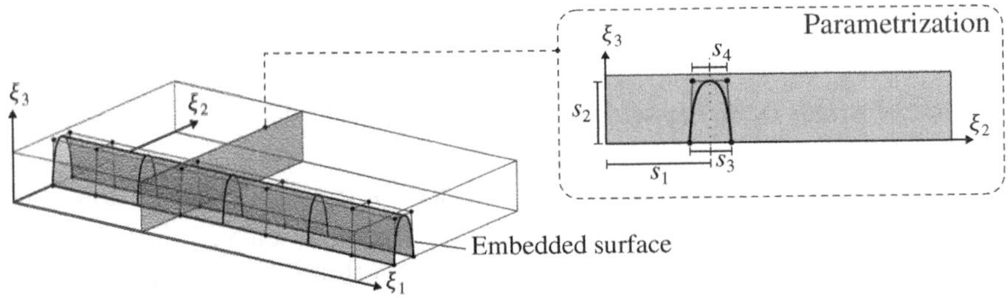

Figure 4.22. *Problem description and geometric construction of the stiffened curved wall. The panel is generated using a quadratic NURBS element. The stiffener is elaborated using a B-spline embedded surface made of 2-by-4 quadratic elements. The design variables are associated with the control points of the embedded surface (four variables for each arch). For a color version of this figure, see www.iste.co.uk/bouclier/IGA.zip*

4.6.2.1. *Settings and geometric modeling*

The problem description, along with its associated geometric construction by the proposed embedded approach, is described in Figure 4.22. The wall originally represents a quarter cylinder (one quadratic NURBS element). We move some of the control points of this initial NURBS surface in order to generate the curved wall and build the volume by the extrusion in the radial direction (see Figure 4.22 (middle), which also provides the coordinates of the control points and their weights). The initial shape of the embedded surface is described in Figure 4.22 (bottom). It consists of a 2-by-4 quadratic B-spline mesh. A set of design variables is defined so as to be able to modify the cross-section and position of the stiffener. With the chosen discretization, six groups of four control points are spread out along the main direction of the stiffener. For each group of control points, we define four design variables, as explained in Figure 4.22 (bottom). Therefore, a total of $6 \times 4 = 24$ design variables are used for this example. The initial shape is obtained by setting the design variables of each group equal to $s_1 = 0.3$, $s_2 = 0.9$, $s_3 = 0.06$ and $s_4 = 0.05$. The area of the stiffener is constrained to be lower than the initial one, which is $A_0 = 21.579$ m^2. Additional inequality constraints between the design variables are included in the optimization problem, in order to prevent the occurrence of undesired shapes during the resolution:

$$s_3 - s_1 \leq 0, \quad s_1 + s_3 - 1 \leq 0, \quad s_4 - s_3 \leq 0.$$

The first two constraints ensure that the embedded surface remains in the parameter space of the volume mapping. The third constraint limits the curvature of the cross-section of the stiffener. Furthermore, additional bound constraints set the range of variation of the design variables. The lower bounds are:

$$s_1^{\text{lo}} = 0., \quad s_2^{\text{lo}} = 0.1, \quad s_3^{\text{lo}} = 0.05, \quad s_4^{\text{lo}} = 0.01,$$

and the upper bounds are:

$$s_1^{\text{up}} = 1., \quad s_2^{\text{up}} = 1., \quad s_3^{\text{up}} = 0.15, \quad s_4^{\text{up}} = 0.15.$$

Finally, the analysis model is defined through the k-refinement of the design model: the wall is discretized with 32-by-32 bi-cubic NURBS (standard Kirchhoff–Love shell) elements and the embedded surface is composed of 16-by-64 cubic B-spline (embedded Kirchhoff–Love shell) elements. The mechanical parameters of the problem are shown in Figure 4.22 (top). The pressure is applied over the mid-surface of the wall. The bottom of the wall and of the stiffener are considered fixed. With this analysis model in hand, the compliance for the initial configuration of the multipatch structure is found to be equal to $C_0 = 38.79$ N m.

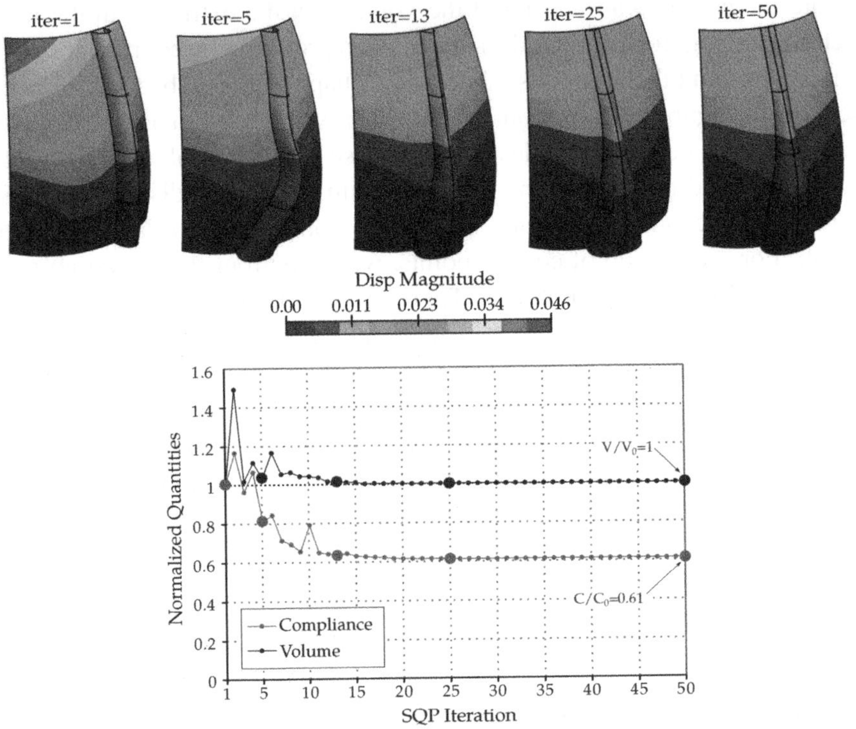

Figure 4.23. *Optimization histories for the resolution of the curved wall problem: (top) evolution of the geometry and of the displacement field, and (bottom) evolution of the compliance and the volume. While sliding, the stiffener perfectly follows the curvature of the wall. For a color version of this figure, see www.iste.co.uk/bouclier/IGA.zip*

4.6.2.2. *Results and discussion*

The optimization results are given in Figures 4.23 and 4.24. During the resolution, the stiffener perfectly follows the curvature of the wall, regardless of its shape complexity. More specifically, the stiffener moves along the wall until it is located in the middle. As for the previous example, this result was predictable given the symmetry of the problem. The optimal shape has a symmetry that demonstrates the good quality of the result. In addition, it may be observed that the cross-section of the stiffener is larger at the bottom than at the top of the wall. This makes sense here since a large cross-section at the bottom improves the fixation of the overall structure. The cross-section also becomes larger where the strains and stresses are critical. The compliance of the final multipatch shape is equal to $C_{opt} = 2.416 \cdot 10^{-3}$, which is 38% lower than for the initial configuration. More details concerning the obtained optimal shape are given in Figure 4.24. In particular, information about the final coordinates of the control points associated with the embedded surface can be found in this figure. Recalling that only local shape variations are performed here, the gain is significant and demonstrates our methodology's ability to handle the local optimization of complex structures. We are not sure whether commercial software (such as Optistruct Altair (n.d.), for example) could handle this type of shape optimization problem, and more specifically, the junction between the stiffener and the wall during the shape modification of the stiffener. Thus, we believe that our framework offers new possibilities from a design point of view: in this context, alternatives to straight and flat stiffeners are possible to further increase the performance of stiffened structures.

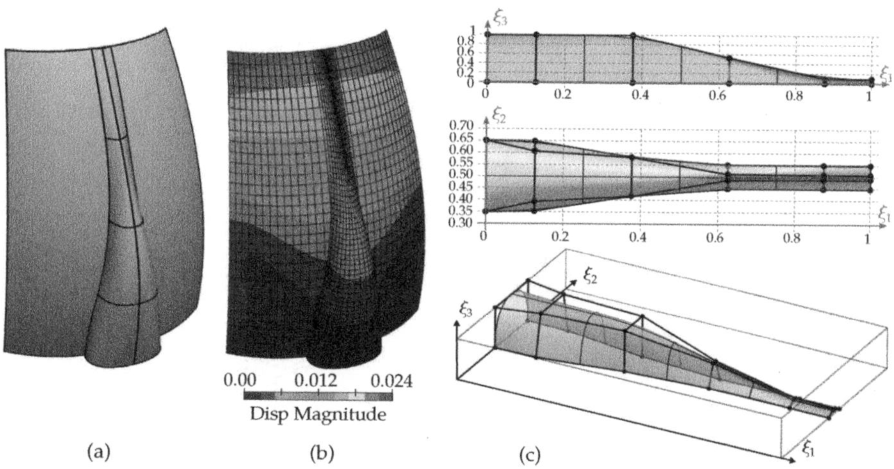

Figure 4.24. *Optimal configuration of the stiffened curved wall: (a) final geometry and (b) the magnitude of the displacement. (c) The final embedded surface: side, top and 3D views. The 2D views give information on the coordinates of the control points. For a color version of this figure, see www.iste.co.uk/bouclier/IGA.zip*

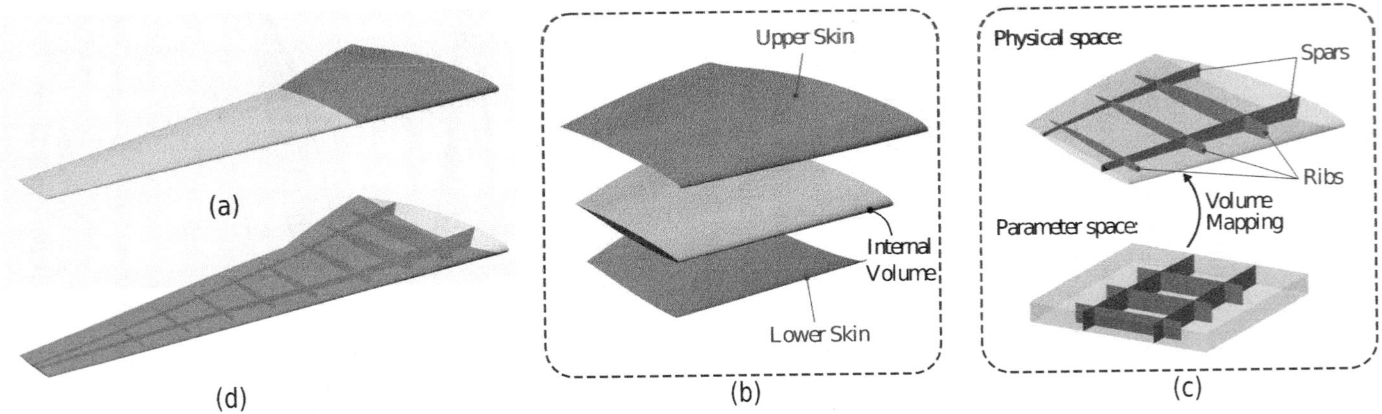

Figure 4.25. *Construction of the model of the wingbox by means of the proposed embedded approach: (a) define the wing skin, (b) generate an internal volume of the wing, (c) embed entities into the parameter space of the created volume and (d) perform volume mapping to obtain the final wing with its internal substructure. For a color version of this figure, see www.iste.co.uk/bouclier/IGA.zip*

4.6.3. *Designing an aircraft wing-box*

With the two examples above, we have mainly shown how the embedded approach enables us to add stiffeners to curved panels. However, the method should not be limited to this purpose. To further underline the flexibility of the developed approach, we now apply it to design the internal substructure of a real wing. Several parameterization-based methods have been recently suggested in order to obtain alternative designs to the classical wing box designs characterized by straight and parallel ribs and spars (Jutte *et al.* 2014; Locatelli *et al.* 2014; Dubois *et al.* 2018). The inspiration for curved structural components comes from dragonfly wings (Jongerius and Lentink 2010). The main issue in these (FE-based) approaches comes from the numerous intermediary (approximation) steps that separate the definition of the design and analysis models. This prevents proper sensitivity analysis and thus makes the use of gradient-based optimization algorithms almost impossible. Furthermore, the geometrical description is limited (most authors start from a planar geometry and then extrude it in the normal direction). Our approach is more generic, and full analytical sensitivities can be computed thanks to the isogeometric link between design and analysis (see section 4.2.4, Hirschler (2019) (Chapters 3 and 5) and Hirschler *et al.* (2020) for additional details regarding isogeometric sensitivity analysis).

4.6.3.1. *Versatile construction of an aircraft wingbox*

We consider the case of a subsonic aircraft wing adapted from Vassberg *et al.* (2008). We build the wing with its internal substructure made of ribs and spars. For this purpose, we adopt the strategy depicted in Figure 4.25.

The starting point is to describe the outer geometry that defines the skin of the wing. This can be done either by using a CAD software, such as Rhinoceros 3D (McNeel & Associates n.d.), or by directly interpolating a point cloud. The NASA common research model (Vassberg *et al.* 2008; Keye *et al.* 2017) is used in this work. The geometric data for the CRM wing (Vassberg *et al.* 2008) enables us to generate a point cloud that represents the outer geometry of the wing. Then, four B-spline surfaces are generated by the interpolation/least-squares approximation of these points. Information about interpolation techniques with splines can be found in Piegl and Tiller (1997). We end up with the surfaces depicted in Figure 4.26.

The next step consists of the construction of a volume that fills the domain delimited by the skins. To generate these volumes, we simply rely on the surfaces. More specifically, we define two B-spline volumes, such that two of their faces match with the upper and lower skins of the wing. The linear degree is chosen in the additional direction of the volume (from the lower surface to the upper one). Thus, no additional control points are generated during the definition of the volumes. An example of a volume is given in Figure 4.27.

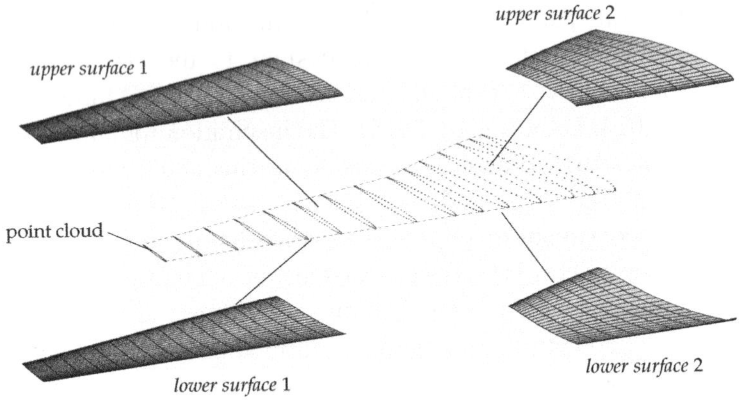

Figure 4.26. *Focus on step 1: defining the outer skin of the wing from a point cloud. Four surfaces are interpolated*

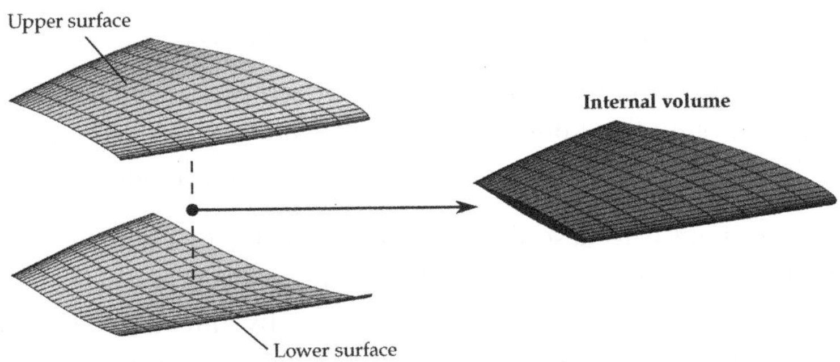

Figure 4.27. *Focus on step 2: the upper and the lower surfaces are used to generate a volume that fills the domain delimited by the skin of the wing*

For each volume, we now have the possibility of embedding spline surfaces in its parameter space, as shown in Figure 4.25(c). The spline composition of the embedded surfaces with volume mapping leads to surfaces that lie perfectly inside of the domain delimited by the skin, and no geometric approximation at the interfaces are introduced (see Figure 4.25(c) and (d)). In the end, it leads to a model with 16 patches (four quadratic standard shell patches for the skin, eight and four quadratic embedded shell patches for the ribs and the spars, respectively). In addition to being tedious to generate, an analysis-suitable model of the wing (i.e. with only matching interfaces, as discussed in section 1.4.2; see, in particular, Figure 1.17) would lead to a much larger number of patches. It could not be achieved without calling upon specific geometric modelers. Here, the modeling task is thus simplified and offers more flexibility.

4.6.3.2. *Structural analysis*

Figure 4.28 presents an example of results that have been obtained on the wing model using the embedded Kirchhoff–Love shell formulation from section 4.4.2 and the DD solver from Chapter 3. The mechanical properties are as follows: each part (the skin, the ribs and the spars) has a Young's modulus $E = 210$ GPa and a Poisson's ratio $\nu = 0.3$. The thickness of the skin is taken as $t^{\text{skin}} = 30$ mm, the thickness of the ribs as $t^{\text{ribs}} = 40$ mm and the thickness of the spars as $t^{\text{spars}} = 50$ mm. The root side of the wing is fixed and an uniform pressure $P = 10$ kPa is applied on the upper skins. For the other geometric settings, we refer, once again to Vassberg *et al.* (2008). A scale factor is applied such that the wing has the following dimensions: the span is 15 m, the sweep angle is 35° and the root and tip chords are equal to 7 m and 1 m, respectively.

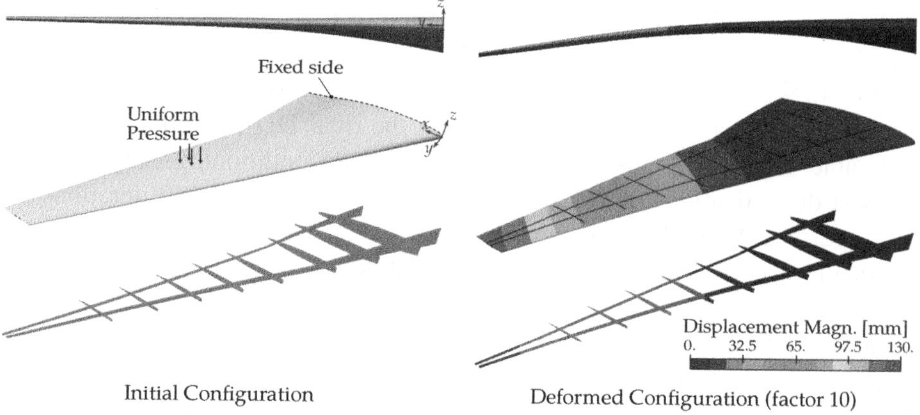

Figure 4.28. *Structural analysis of the CRM wing (global view). For a color version of this figure, see www.iste.co.uk/bouclier/IGA.zip*

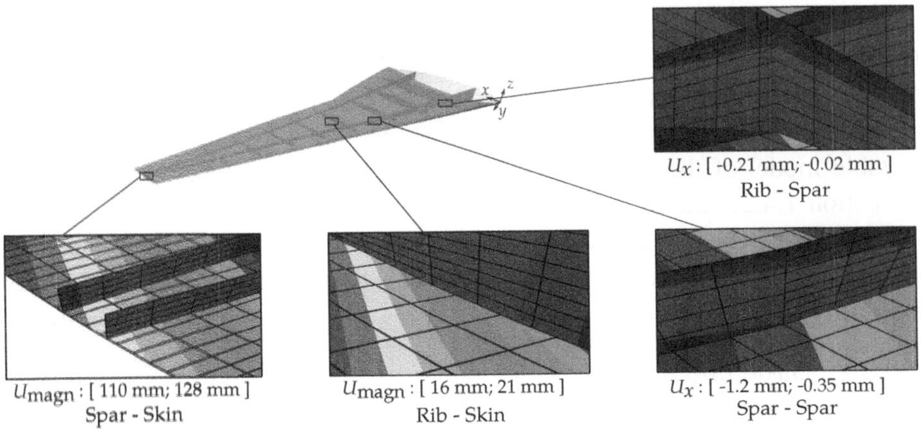

Figure 4.29. *Structural analysis of the CRM wing (local views). For a color version of this figure, see www.iste.co.uk/bouclier/IGA.zip*

	Wing			
Ref	N_{tot}	N_λ	**I**	$\tilde{\mathbf{F}}_A^{-1}$
0	9,246	2,272	1,874	258
1	28,554	4,096	-	299
2	102,444	7,744	-	378

Table 4.1. *Performance study for the wing problem*

In order to highlight the correct imposition of the coupling conditions, we depict detailed views around the different non-conforming interfaces in Figure 4.29. We can see the continuity of the displacement over these interfaces. Concerning the convergence of the dual DD algorithm, we were only able to run the resolution without the preconditioning step for the coarsest discretization. In this case, it already took almost 2,000 iterations to converge (see Table 4.1). However, the generalized Dirichlet preconditioner drastically accelerated the convergence, as shown in Table 4.1. It enabled us to analyze refined versions of the wing model in an acceptable number of iterations. We obtained the slow growth of the iteration count versus the refinement level. It demonstrates that the generalized Dirichlet preconditioner introduced in section 3.4.4 is robust, even for more complex cases, such as the aircraft wing.

• Geometric Modeling

• Shape parametrization

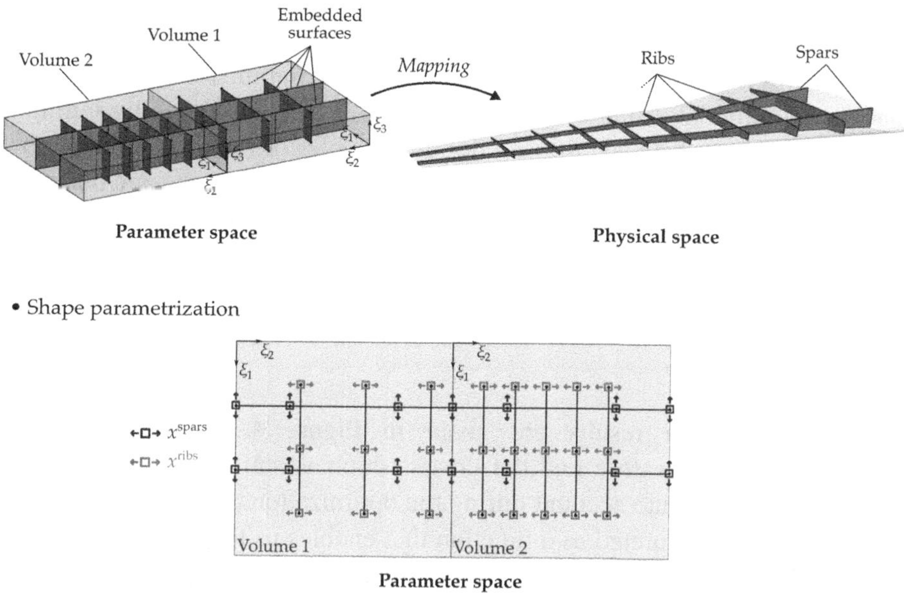

Figure 4.30. *Construction of the geometric model of the representative wing (top) and definition of the design variables for the shape optimization of its internal substructure (bottom)*

4.6.3.3. *Shape optimization*

As in the previous example, here, we perform local optimization, i.e. we move the control points associated with the embedded surfaces to impose shape modifications to the internal substructure of the wing. Once again, we minimize compliance under the area constraint of $A \leq A_0$. The geometric model and the definition of the design variables are depicted in Figure 4.30. Each rib is described with a B-spline of one single element, with degree 2 in the main direction, and degree 1 in the vertical direction. The same applies for the spars, except they are discretized with two elements, which makes it possible to set more design parameters. The control points of the spars can move in the ξ_1-direction of the volume parameter space (see Figure 4.30 (bottom)). The same motion is prescribed to the edges joining the upper and lower skin. Thus, four design variables per spar are set. One common design variable is defined at the junction between the spars embedded in volumes 1 and 2, in order to maintain their connection. In addition, three design variables per rib are defined, and move the control points in the ξ_2-direction of the volume parameter space (see Figure 4.30 (bottom)). We end up with a total of 38 design variables. We also add geometric constraints into the formulation of the optimization problem that guarantee a minimal distance between the ribs. It is simply done by setting a minimal difference between the two design variables of two neighboring ribs. We also voluntarily skip to enforce the coupling between the ribs and the spars for the sake of simplicity. In fact, it avoids locating their intersections during the optimization process. The wing is finally considered to be fixed on its right end, and pressure is applied to its top skin. The geometric and mechanical model parameters are the same as for the structural analysis carried out above (see Figure 4.28).

The optimization results are given in Figure 4.31. Thanks to the construction, the curvature and the cross-section variations of the wing are automatically taken into account during the optimization process. The volume mapping can be interpreted as a function that enables us to keep the structural components inside of the volume delimited by the skins. No additional procedures that could have broken the link between the shape parameters and the final geometric model of the wing are required, as in alternative FE-based approaches (Jutte *et al.* 2014; Locatelli *et al.* 2014; Dubois *et al.* 2018). We can observe the evolution of the deformation of the wing during the optimization. The overall displacement of the wing is drastically reduced. The maximal displacement of the final design is about 10 times lower than the maximal displacement of the initial design (129 mm for the initial, versus 12.8 mm for the optimal). The compliance is therefore reduced to a great extent, $C^{\text{opt}}/C_0 = 0.213$, whereas the overall area is kept constant. The final design has ribs and spars with curved shapes. It proves that curved structural components can largely improve the mechanical behavior of the wing. It coincides with the observations from other recent works (Mulani *et al.* 2013; Jutte *et al.* 2014; Locatelli *et al.* 2014; Zhao and Kapania 2016; Dubois *et al.* 2018; Wang *et al.* 2019). Such innovative designs can surely improve further aircraft structures and, more generally, stiffened structures. It also motivates the construction of a robust and efficient framework that enables us to design these new future designs. We believe that such numerical experiments account for our method's potential to meet this challenge.

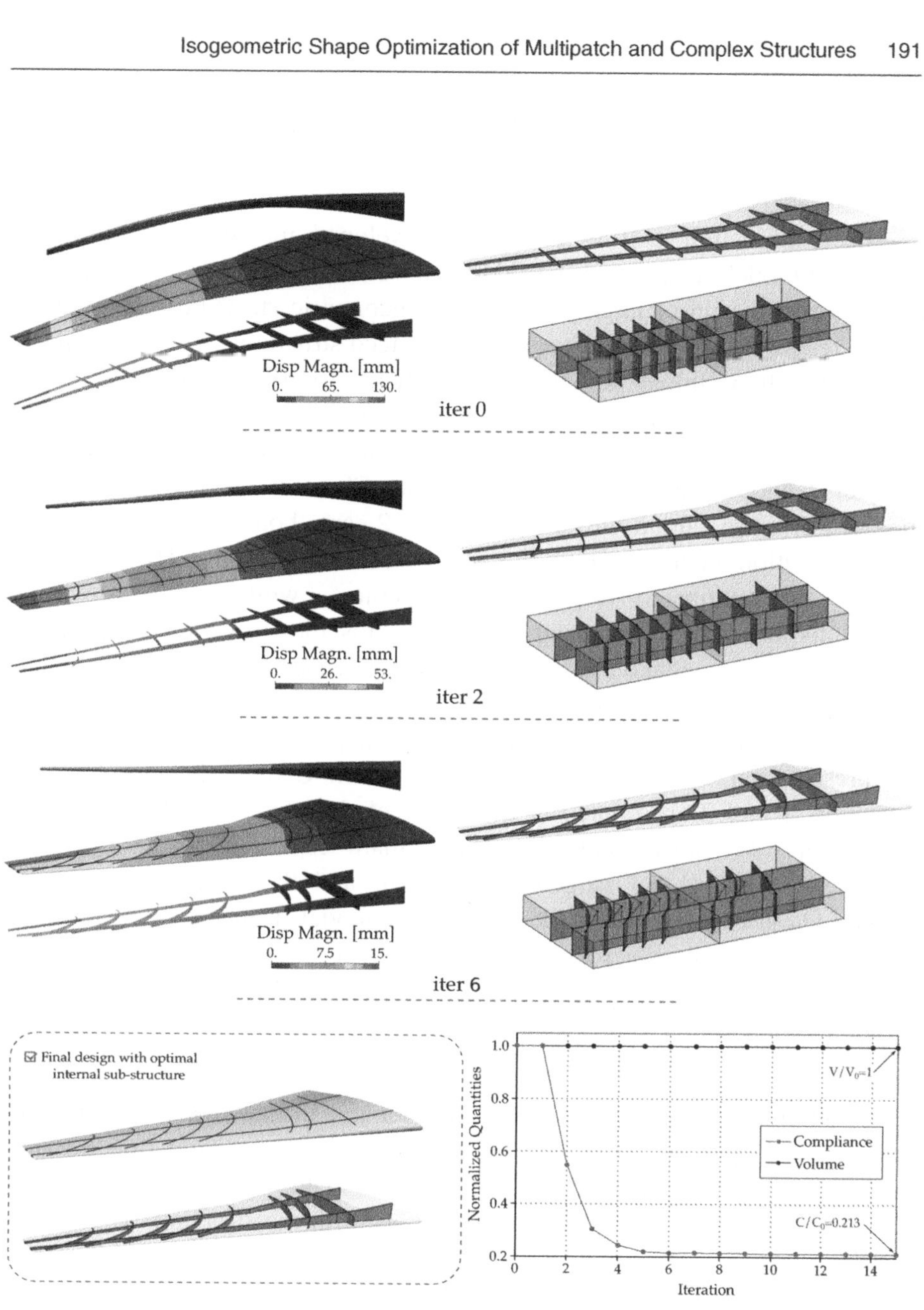

Figure 4.31. *Optimization histories for the resolution of the wing problem, and final design of the ribs and spars. For a color version of this figure, see www.iste.co.uk/bouclier/IGA.zip*

4.7. Towards the optimal design of structural details within isogeometric patches

The combination of the DD solver developed in Chapter 3 with the embedded modeling approach introduced in section 4.4, allows for the proper and effective design of large multipatch isogeometric structures. To go even further regarding the complexity of the envisaged structures, we now open the door, through a simple numerical experiment, for the efficient design of structural details within patches of large macro-structures. To this end, the idea here is to make use of the modeling and analysis strategies built in Chapter 2. Indeed, as highlighted throughout Chapter 3, the DD solver and the non-invasive algorithms of Chapter 2 are both appropriate in a multi-resolution context. Only the application differs: the DD solver is suitable for multipatch isogeometric structures, while the non-invasive algorithms are adapted to integrate local details into isogeometric patches. More precisely, we make use of the newly developed non-invasive algorithm [2.29]–[2.30] to adapt to any non-conforming spline-based configuration that may be encountered through an isogeometric shape optimization process. For the presentation, we recycle the notations from Chapter 2; in particular, subscripts 1 and 2 are related to quantities of the global and local models, respectively. We also recall that the domain $\Omega_1 = \Omega_{11} \cup \Omega_{12} \cup \Gamma$ characterizes the global isogeometric patch, and the local model of the domain Ω_2 is expected to replace the global model in the sub-domain Ω_{12}, through data exchanges across Γ (see Figure 2.1 as a reminder).

4.7.1. *A simple but instructive test case*

As an enlightening example, we consider the simple shape optimization problem described in Figure 4.32(a). The test case comes from Bouclier and Passieux (2018). The position of an inclusion within a global 2D plate is optimized so as to provide minimal compliance for the structure. The design variables are the horizontal and vertical coordinates of the center of the inclusion, denoted by x_c and y_c in the physical space, respectively (see Figure 4.32(a)). The inclusion is constrained to fully stay within the plate. Standard 2D elasticity (plane stress assumption) is used for the mechanical modeling. The Young's modulus of the inclusion is chosen to be a hundred times smaller than that of the plate ($E_1 = 10^3$ and $E_2 = 10$). The other model parameters are taken to be similar to that in the plate, with the central inclusion example from Chapter 2 (see Figure 2.12(a)): horizontal length of the plate $L = 7$, vertical length of the plate $H = 10$, radius of the inclusion $R_i = 1$, Poisson's ratios $\nu_1 = \nu_2 = 0.3$ and in-plane tension $p = 10$. Regarding the discretization, a regular quadratic B-spline grid composed of 28-by-40 elements is used for the plate, while a circular quadratic NURBS mesh with 48- (circumferential direction) by-12 (radial direction) elements is constructed to model the inclusion.

As an optimization algorithm, we make use, this time, of the black box `fminsearch` available in `matlab`. This routine uses a gradient-free algorithm: it is the Nelder–Mead simplex algorithm described in Lagarias *et al.* (1998). More sophisticated iterative schemes, like those based on the gradient as in previous sections, could be used to improve the optimization performance. Nevertheless, as our goal is to prove the interest of making use of the non-invasive solver developed in Chapter 2 (section 2.4) for the analysis step (see Figure 4.1, step 2), rather than building an advanced optimization tool, we choose to initiate the study with this simple algorithm. Starting with an inclusion located at position $x_c^0 = 2$ and $y_c^0 = 2$, the optimization process is expected to lead to the configuration of a central inclusion, i.e. such that $x_c^{ref} = 3.5$ and $y_c^{ref} = 5$. Indeed, the further the inclusion is from all plate boundaries, the less stress concentration is encountered around the inclusion. For illustration purposes, the solution, in terms of von Mises stress, is plotted for the initial configuration and the expected optimized one (see Figure 4.32(b) and 4.32(c), respectively).

4.7.2. *Unify the non-invasive global/local approach and the optimization of local details*

The optimization algorithm calls upon the non-conforming, non-invasive algorithm [2.29]-[2.30] to get the displacement solution at each configuration encountered through the optimization process. A maximum of 20 iterations is prescribed for the non-invasive iterative algorithm. From a general point of view, we recover the benefits that are close to those of the DD solver from Chapter 3 for the shape optimization of multipatch structures (see section 4.3): (i) no need for cumbersome meshing procedures to impose matching (or even non-matching) meshes between the plate and the inclusion, (ii) no need to fully rebuild and recompute the global/local structure at each optimization iteration and (iii) possibility of taking the solution of the previous design n as an initial guess to speed up the non-invasive solver running at optimization step $n+1$.

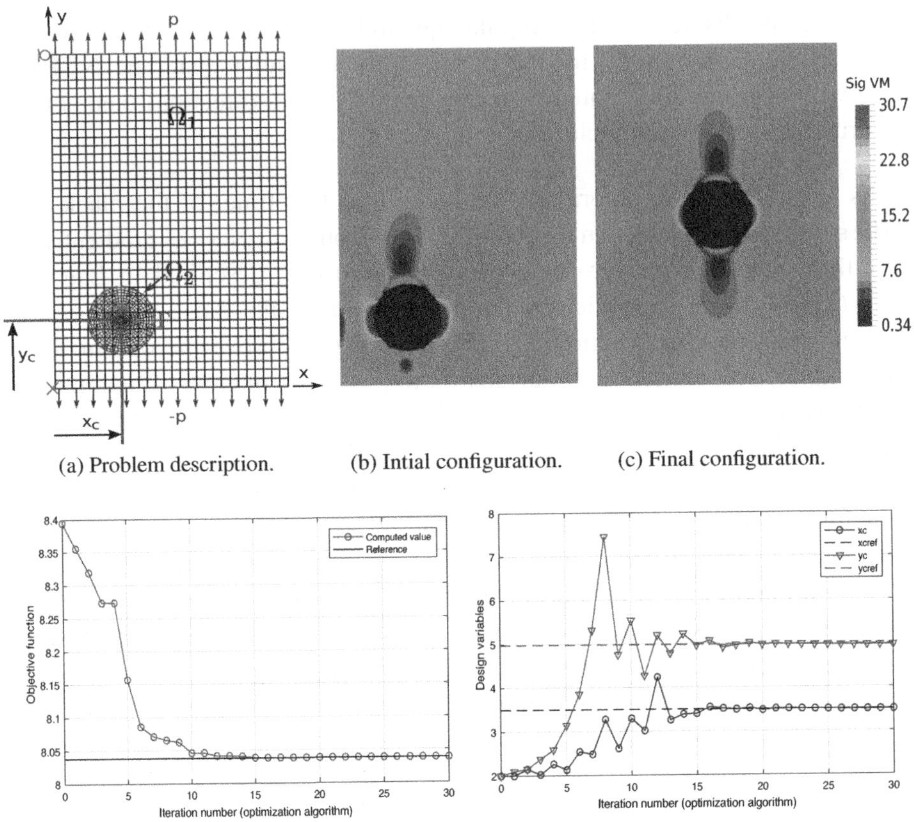

(a) Problem description. (b) Intial configuration. (c) Final configuration.

(d) Objective functions w.r.t. iterations. (e) Design variables w.r.t. iterations.

Figure 4.32. *Simple optimization problem: the position of the inclusion* (x_c, y_c) *in the 2D plate is optimized so as to provide minimal compliance for the structure. The optimization algorithm calls upon the non-conforming non-invasive Nitsche solver from Chapter 2 (section 2.4) each time the objective function needs to be evaluated. This simplifies the optimization process, provides more robustness and offers computational time saving. For a color version of this figure, see www.iste.co.uk/bouclier/IGA.zip*

More precisely, focusing on point (ii) in the situation described in Figure 4.32(a), the stiffness operators $\mathbf{K}_1$ and $\left[\mathbf{K}_2 + \left(\overline{\mathbf{K}}_{22}^{N\,T} - \overline{\mathbf{K}}_{22}^{N}\right)\right]$ in [2.29]–[2.30] are assembled and factorized once for all in a pre-processing step. Then, the only things that need to be done to compute the successive coupled solutions encountered at each iteration of the optimization process are:

1) build the integration rule for the small immersed region Ω_{12};

2) assemble $\mathbf{r}_{12}$, $\overline{\mathbf{K}}_{11}^{N}$, $\overline{\mathbf{K}}_{12}^{N}$ and $\overline{\mathbf{K}}_{21}^{N}$;

3) perform a few lower and upper triangular resolutions to solve problems [2.29] and [2.30] alternatively.

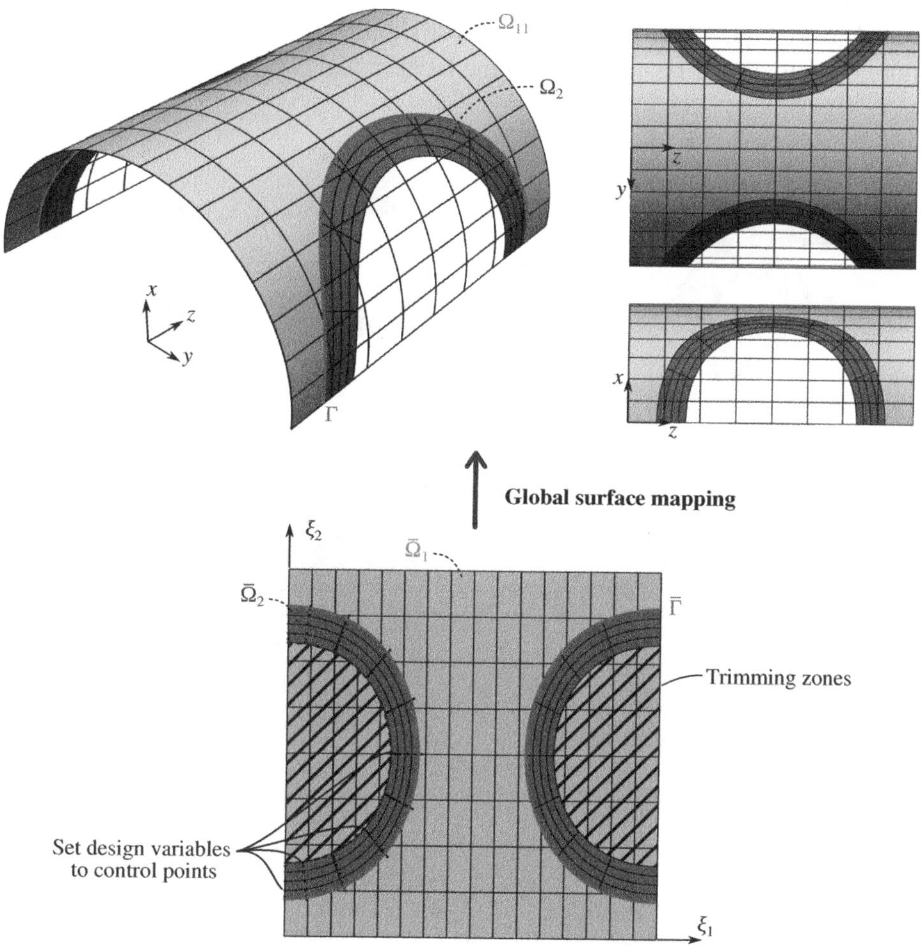

Figure 4.33. *Application of the embedded shape update approach for designing geometrical details (holes) within a cylindrical shell. The design variables move the control points of the trimming curve defining the boundary of the holes in the parameter space of the shell. For a color version of this figure, see www.iste.co.uk/bouclier/IGA.zip*

Conversely, a direct solver would require the resulting global/local problem [2.23]–[2.24] (that may also be ill-conditioned) to be fully rebuilt and recomputed at each of these steps.

NOTE.– It may be noted that in the more general situation where the shape of the local detail is modified (along with its position), the local operator $\left[\mathbf{K}_2 + \left(\overline{\mathbf{K}}_{22}^{N\,T} - \overline{\mathbf{K}}_{22}^{N}\right)\right]$ needs to be updated at each optimization iteration as well. However, we still save the assembling and resolution of the global model (over Ω_1), which is expected to be significantly larger (thus requiring more computational resources) than the local model over Ω_2.

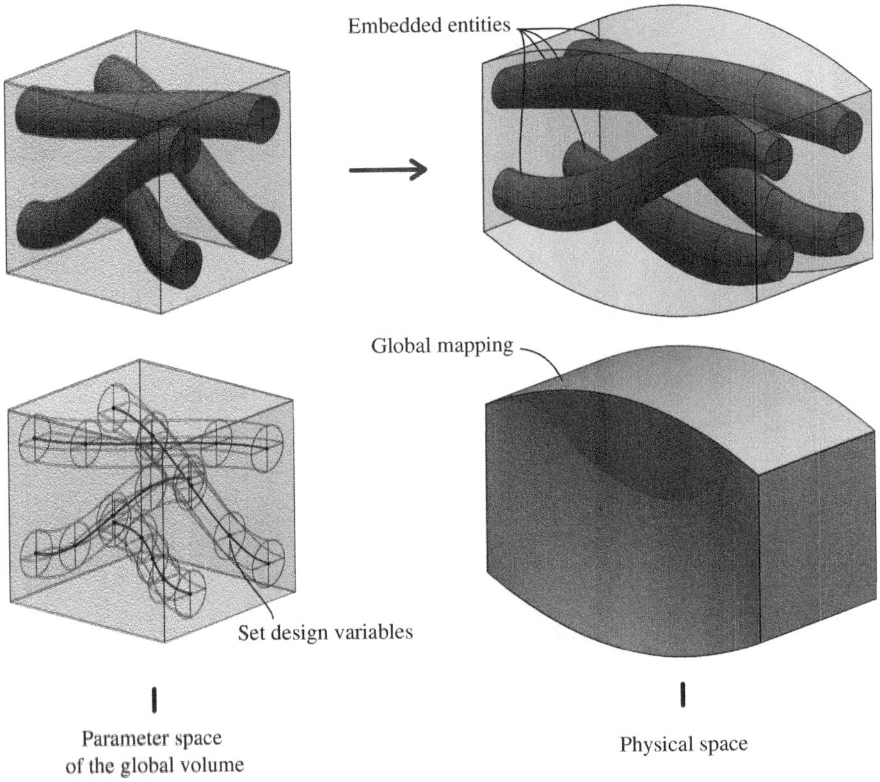

Figure 4.34. *Application of the embedded shape update approach for designing geometrical details (fibers) within a composite. The design variables move the control points of the fibers (let us say, only the corresponding mid-lines) in the parameter space of the volume defining the matrix of the composite*

4.7.3. *Preliminary results and perspectives*

Figure 4.32(d) and (e) shows the convergence of the objective function and of the design variables, respectively. The convergence criterion for the optimization algorithm relies on the stagnation of the objective function, as well as of the design variables: it stops when it gets below 10^{-4}. It can be observed that we are able to recover the expected optimized configuration through the optimization process, which accounts for the ability of the non-invasive solver to adapt to any arbitrary non-conforming spline situations, in particular when the edge of an element of the local model approaches that of an element of the global model.

Even if further studies have not been carried out yet, we believe that the use of a dedicated global/local non-invasive algorithm should definitely facilitate the process and ensure computational time saving in the context of shape optimization of structural details within isogeometric patches. In addition, let us point out that the strategy may be easily applied to complex shell and solid structures, such as the ones of previous sections. In this case, an analog treatment, such as in Figure 4.32(a), for the design modification could be performed in the parameter space of the shell or the volume. This would lead to an embedded-like shape update strategy according to the terminology introduced in this chapter (see, for example, Figures 4.7 (bottom) and 4.9 (bottom)). Illustrations are provided in the case of a shell and of a solid structure in Figures 4.33 and 4.34, respectively. The spline mapping of the considered structure would, in some sense, act as the volume mapping [4.29] to transfer the shape of the local detail to the physical space, while ensuring that this detail is perfectly incorporated into the true structure. This strategy would enable us to optimize complex trimmed structures, such as in the example of Figure 4.33 inspired by Kang and Youn (2016), where the shape of two holes within a shell would be designed. With Figure 4.34, we also see that the strategy could be applied to finer scales for designing optimal micro-structured geometries, such as the ones of composite materials (Desmorat 2002; Chelaghma 2018; Naylor 2019) possibly coming from additive manufacturing (Massarwi *et al.* 2018; Antolin *et al.* 2019; Li *et al.* 2020; Papapetrou *et al.* 2020; Yera *et al.* 2020). From this point of view, the success of the study shown in Figure 4.32 and the combination of this result with the results from previous sections open the door for the development of effective optimization tools, based on some non-invasive and DD numerical algorithms, to handle realistic applications. This motivates future research in this direction and, more generally, encourages us to make use of such algorithms when performing multi-resolution processes.

4.8. Summary and discussion

This final chapter addresses the optimal design of complex structures in the context of IGA. We recall that the cornerstone of the problem concerns the handling of multipatch models with non-conforming parameterizations. After answering the issue from an analysis point of view with the construction of dedicated DD solvers in Chapter 3, we first highlighted the interests in terms of the efficiency and flexibility of integrating such algorithms in shape optimization loops. The parallel architecture is indeed completely adapted to compute the response functions and sensitivities of multipatch structures, and allows for great computational time savings, in particular in the case of non-design areas and when performing smart re-analysis. Then, the chapter is largely devoted to the construction of a full geometrical framework that enables us to transform a complex multipatch model with non-conforming interfaces into what we may call an *optimization-suitable* model, in the sense that they are ready for natural isogeometric shape optimization.

We start our development by focusing on shells in view of being able to tackle representative aerostructures, thus including many stiffened panels. More precisely, by drawing inspiration from the free-form deformation (FFD) concept (Sederberg and Parry 1986; Kenway and Martins 2014; Bletzinger 2017), we introduce an embedded geometrical modeling that enables us to properly impose shape changes to stiffened structures. It simply consists of defining the stiffener mid-surface, using a spline composition of an embedded surface with a volume mapping related to the panel. Using this embedded (or immersed) approach, we can either modify the embedded surface for local optimization or the volume mapping for global optimization of complex shell structures. Of course, this approach is not restricted to stiffened structures and results are also provided for general multipatch shells. The final geometric model obtained by spline composition is then directly used in the analysis step by formulating a new isogeometric, embedded Kirchhoff–Love shell element. This means that the original idea behind IGA is maintained, since both the design and analysis are performed on a common geometrical model. It is particularly attractive in the context of structural optimization, since it provides an exact link between the shape parameters and the resulting analysis model, thus allowing us to derive full analytical sensitivities for accurate and effective gradient-based optimization. Our approach is thus very robust and generic in contrast to the current FE-based framework (Jutte *et al.* 2014; Locatelli *et al.* 2014; Dubois *et al.* 2018), where several intermediary approximation steps separate the design and analysis models. We have been able, in particular, to create a versatile model of a wing with its internal substructure. Due to the use of embedded entities, the internal substructure automatically follows the outer geometry of the wing during shape update. This brings about the possibility of exploring new types of design while in usual aerostructural optimization and sizing variables, such as the thicknesses of the shell components, are included in the design process (Kenway and Martins 2014; Keye *et al.* 2017).

The generalization of the methodology to complex solid models is also outlined. The strategy is fully consistent with innovative practices in the field of geometrical modeling, which consists of using functional composition (Elber 2017) to generate complex spline models, such as those encountered in additive manufacturing (Massarwi *et al.* 2018; Antolin *et al.* 2019). The developed approach thus seems relevant for performing the shape optimization of such complex trivariate structures. At this stage, we recall that the fundamental ingredient is to make use of real spline compositions (in the sense that the composition is performed over the whole embedded domain). This is in contrast to the standard practice in (FE-based) FFD shape optimization (Lyu and Martins 2015; Bletzinger 2017), where the morphing box only acts on the FE nodes (or control points to a certain extend), which would fail to impose shape updates that naturally guarantee the inter-patch geometric continuity.

To go even further regarding the complexity of the envisaged structures, we open the door, by carrying out a simple numerical experiment, for the efficient design of structural details within large macro-structures. The idea is to make use, this time, of the global/local non-invasive strategies from Chapter 2 for the analysis step within the optimization process. From a general point of view, we recover similar benefits as when using the DD solvers from Chapter 3 for the shape optimization of multipatch structures, namely it avoids calling upon cumbersome meshing procedures and rebuilding and recomputing the whole coupled problem at each optimization iteration. This should, for example, foster the handling of trimmed isogeometric entities, thus offering a simple process to conduct the shape optimization of geometrical details (such as holes (Wall *et al.* 2008; Fußeder *et al.* 2015) or inclusions), or even to perform a local topology optimization (Seo *et al.* 2010; Parvizian *et al.* 2012; Groen *et al.* 2017).

Overall, given its robustness and versatility, the developed framework may be of interest in many engineering fields, such as multidisciplinary design optimization (MDO) (Balesdent *et al.* 2012; Martins and Lambe 2013a); indeed, we could simultaneously optimize the outer geometry of the aerostructure regarding some aerodynamic criteria and the shape of the internal substructure (possibly including even more local details) regarding the structural behavior. Other prospects of our methodology concern the extension to more sophisticated structural behaviors, for example, involving vibration or buckling, which are crucial for slender structures (Lund and Stegmann 2006; Nagy 2011; Stanford *et al.* 2014). If the derivation of analytical sensitivities in the situation of eigenfrequency optimization seems straightforward, it may be easier to resort to some solid-shell formulations (Bouclier *et al.* 2015b; Caseiro *et al.* 2015) in the case of buckling optimization. Perhaps the design of dedicated optimization algorithms would be necessary here. Once again, note that the proposed approach is surely not limited to aerostructures or even shells, but could basically be applied to any other models composed of global and local parts (one can think, for example, of designing optimal micro-structured geometries using a similar procedure (Massarwi *et al.* 2018; Antolin *et al.* 2019)). According to us, the use of the FFD concept in a continuous way to perform isogeometric shape optimization is a relevant idea and should be widely considered when the objective is to address complex situations. It can actually be viewed as the counterpart of the immersed (or embedded) approach for numerical simulation (Burman *et al.* 2015; Schillinger and Ruess 2015; Elfverson *et al.* 2019). FFD-based shape optimization offers the opportunity to decouple the design space from the actual geometry, while immersed methods allow the approximation of the displacement to be distinguished from the discretization of the geometry. The same attractiveness as the one attributed to immersed methods in the context of scientific computing is thus expected when using the FFD concept in the field of shape optimization.

References

Adam, C., Hughes, T.J.R., Bouabdallah, S., Zarroug, M., Maitournam, H. (2015). Selective and reduced numerical integrations for NURBS-based isogeometric analysis. *Computer Methods in Applied Mechanics and Engineering*, 284, 732–761.

Akhras, H.-A., Elguedj, T., Gravouil, A., Rochette, M. (2016). Isogeometric analysis-suitable trivariate NURBS models from standard B-Rep models. *Computer Methods in Applied Mechanics and Engineering*, 307, 256–274.

Akkerman, I., Bazilevs, Y., Kees, C.E., Farthing, M.W. (2011). Isogeometric analysis of free-surface flow. *Journal of Computational Physics*, 230(11), 4137–4152.

Altair (n.d.). OptiStruct [Online]. Available at: https://www.altair.com/optistruct/.

Ambati, M., Kiendl, J., De Lorenzis, L. (2018). Isogeometric Kirchhoff–Love shell formulation for elasto-plasticity. *Computer Methods in Applied Mechanics and Engineering*, 340, 320–339.

Amini, A.M., Dureisseix, D., Cartraud, P., Buannic, N. (2009). A domain decomposition method for problems with structural heterogeneities on the interface: Application to a passenger ship. *Computer Methods in Applied Mechanics and Engineering*, 198(41), 3452–3463.

Annavarapu, C., Hautefeuille, M., Dolbow, J. (2012). A robust Nitsche's formulation for interface problems. *Computer Methods in Applied Mechanics and Engineering*, 225, 44–54.

Antolin, P., Buffa, A., Fabre, M. (2018). *A priori* error for unilateral contact problems with Lagrange multipliers and isogeometric analysis. *IMA Journal of Numerical Analysis*, 39(4), 1627–1651.

Antolin, P., Buffa, A., Cohen, E., Dannenhoffer, J.F., Elber, G., Elgeti, S., Haimes, R., Riesenfeld, R. (2019). Optimizing micro-tiles in micro-structures as a design paradigm. *Computer-aided Design*, 115, 23–33.

Apostolatos, A., Breitenberger, M., Wüchner, R., Bletzinger, K.-U. (2014a). Domain decomposition methods and Kirchhoff–Love shell multipatch coupling in isogeometric analysis. In *Isogeometric Analysis and Applications 2014*, Jüttler, B. and Simeon, B. (eds). Springer International Publishing.

Apostolatos, A., Schmidt, R., Wüchner, R., Bletzinger, K.-U. (2014b). A Nitsche-type formulation and comparison of the most common domain decomposition methods in isogeometric analysis. *International Journal for Numerical Methods in Engineering*, 97, 473–504.

Apostolatos, A., De Nayer, G., Bletzinger, K.-U., Breuer, M., Wüchner, R. (2019). Systematic evaluation of the interface description for fluid–structure interaction simulations using the isogeometric mortar-based mapping. *Journal of Fluids and Structures*, 86, 368–399.

Arnold, D., Brezzi, F., Cockburn, B., Marini, D. (2002). Unified analysis of discontinuous Galerkin methods for elliptic problems. *SIAM Journal of Numerical Analysis*, 39, 1749–1779.

Auricchio, F., Da Veiga, L.B., Hughes, T.J.R., Reali, A., Sangalli, G. (2010). Isogeometric collocation methods. *Mathematical Models and Methods in Applied Sciences*, 20(11), 2075–2107.

Auricchio, F., Calabro, F., Hughes, T.J.R., Reali, A., Sangalli, G. (2012a). A simple algorithm for obtaining nearly optimal quadrature rules for NURBS-based isogeometric analysis. *Computer Methods in Applied Mechanics and Engineering*, 249, 15–27.

Auricchio, F., Da Veiga, L.B., Hughes, T.J.R., Reali, A., Sangalli, G. (2012b). Isogeometric collocation for elastostatics and explicit dynamics. *Computer Methods in Applied Mechanics and Engineering*, 249, 2–14.

Badia, S. and Verdugo, F. (2018). Robust and scalable domain decomposition solvers for unfitted finite element methods. *Journal of Computational and Applied Mathematics*, 344, 740–759.

Baiges, J., Codina, R., Henke, F., Shahmiri, S., Wall, W. (2012). A symmetric method for weakly imposing Dirichlet boundary conditions in embedded finite element meshes. *International Journal for Numerical Methods in Engineering*, 90, 636–658.

Balesdent, M., Bérend, N., Dépincé, P., Chriette, A. (2011). A survey of multidisciplinary design optimization methods in launch vehicle design. *Structural and Multidisciplinary Optimization*, 45(5), 619–642.

Balesdent, M., Bérend, N., Dépincé, P., Chriette, A. (2012). A survey of multidisciplinary design optimization methods in launch vehicle design. *Structural and Multidisciplinary Optimization*, 45(5), 619–642.

Bandara, K. and Cirak, F. (2017). Isogeometric shape optimisation of shell structures using multiresolution subdivision surfaces. *Computer-aided Design*, 95, 62–71.

Bauer, A., Breitenberger, M., Philipp, B., Wüchner, R., Bletzinger, K.-U. (2017). Embedded structural entities in NURBS-based isogeometric analysis. *Computer Methods in Applied Mechanics and Engineering*, 325, 198–218.

Bazilevs, Y., Calo, V., Cottrell, J., Hughes, T.J.R., Reali, A., Scovazzi, G. (2007). Variational multiscale residual-based turbulence modeling for large eddy simulation of incompressible flows. *Computer Methods in Applied Mechanics and Engineering*, 197(1–4), 173–201.

Bazilevs, Y., Calo, V., Cottrell, J.-A., Evans, J.-A., Hughes, T.J.R., Lipton, S., Scott, M., Sederberg, T. (2010). Isogeometric analysis using T-splines. *Computer Methods in Applied Mechanics and Engineering*, 199, 229–263.

Bazilevs, Y., Hsu, M.-C., Scott, M. (2012). Isogeometric fluid–structure interaction analysis with emphasis on non-matching discretizations, and with application to wind turbines. *Computer Methods in Applied Mechanics and Engineering*, 249, 28–41.

Béchet, É., Moës, N., Wohlmuth, B. (2009). A stable Lagrange multiplier space for stiff interface conditions within the extended finite element method. *International Journal for Numerical Methods in Engineering*, 78(8), 931–954.

Belytschko, T., Stolarski, H., Liu, W.K., Carpenter, N., Ong, J.S. (1985). Stress projection for membrane and shear locking in shell finite elements. *Computer Methods in Applied Mechanics and Engineering*, 51(1–3), 221–258.

Ben Belgacem, F. and Maday, Y. (1997). The mortar element method for three dimensional finite elements. *ESAIM: Mathematical Modelling and Numerical Analysis – Modélisation mathématique et analyse numérique*, 31(2), 289–302.

Benson, D., Bazilevs, Y., Hsu, M., Hughes, T.J.R. (2010). Isogeometric shell analysis: The Reissner–Mindlin shell. *Computer Methods in Applied Mechanics and Engineering*, 199(5), 276–289.

Bernadou, M. and Cubier, A. (1998). Numerical analysis of junctions between thin shells. Part 1: Continuous problems. *Computer Methods in Applied Mechanics and Engineering*, 161, 349–363.

Bernadou, M., Fayolle, S., Léné, F. (1989). Numerical analysis of junctions between plates. *Computer Methods in Applied Mechanics and Engineering*, 74, 307–326.

Bettinotti, O., Allix, O., Perego, U., Oancea, V., Malherbe, B. (2014). A fast weakly intrusive multiscale method in explicit dynamics. *International Journal for Numerical Methods in Engineering*, 100(8), 577–595.

Bischoff, M., Ramm, E., Irslinger, J. (2017). Models and finite elements for thin-walled structures. In *Encyclopedia of Computational Mechanics, Part 1: Solids and Structures*, 2nd edition. John Wiley & Sons, New York.

Blanchard, M., Allix, O., Gosselet, P., Desmeure, G. (2019). Space/time global/local noninvasive coupling strategy: Application to viscoplastic structures. *Finite Elements in Analysis and Design*, 156, 1–12.

Bletzinger, K.-U. (2014). A consistent frame for sensitivity filtering and the vertex assigned morphing of optimal shape. *Structural and Multidisciplinary Optimization*, 49(6), 873–895.

Bletzinger, K.-U. (2017). Shape optimization. In *Encyclopedia of Computational Mechanics, Part 2: Solids and Structures*. John Wiley & Sons, New York.

Bletzinger, K.-U., Kimmich, S., Ramm, E. (1992). Efficient modeling in shape optimal design. *Computing Systems in Engineering*, 2, 483–495.

Bletzinger, K.-U., Wüchner, R., Daoud, F., Camprubí, N. (2005). Computational methods for form finding and optimization of shells and membranes. *Computer Methods in Applied Mechanics and Engineering*, 194(30–33), 3438–3452.

Bletzinger, K.-U., Firl, M., Daoud, F. (2008). Approximation of derivatives in semi-analytical structural optimization. *Computers & Structures*, 86(13–14), 1404–1416.

Bletzinger, K.-U., Firl, M., Linhard, J., Wüchner, R. (2010). Optimal shapes of mechanically motivated surfaces. *Computer Methods in Applied Mechanics and Engineering*, 199(5), 324–333.

Bouclier, R. (2020). Some numerical tools based on IGA for optimization applications in structural mechanics. HDR thesis, Université fédérale de Toulouse-Midi-Pyrénées.

Bouclier, R. and Passieux, J.-C. (2018). A Nitsche-based non-intrusive coupling strategy for global/local isogeometric structural analysis. *Computer Methods in Applied Mechanics and Engineering*, 340, 253–277.

Bouclier, R. and Passieux, J.-C. (2022). *IGA: Non-invasive Coupling with FEM and Regularization of Digital Image Correlation Problems*. ISTE Ltd, London and John Wiley & Sons, New York.

Bouclier, R., Elguedj, T., Combescure, A. (2012). Locking free isogeometric formulations of curved thick beams. *Computer Methods in Applied Mechanics and Engineering*, 245, 144–162.

Bouclier, R., Elguedj, T., Combescure, A. (2013a). Efficient isogeometric NURBS-based solid-shell elements: Mixed formulation and B-bar method. *Computer Methods in Applied Mechanics and Engineering*, 267, 86–110.

Bouclier, R., Elguedj, T., Combescure, A. (2013b). On the development of NURBS-based isogeometric solid shell elements: 2D problems and preliminary extension to 3D. *Computational Mechanics*, 52(5), 1085–1112.

Bouclier, R., Elguedj, T., Combescure, A. (2015a). Development of a mixed displacement-stress formulation for the analysis of elastoplastic structures under small strains: Application to a locking-free, NURBS-based solid-shell element. *Computer Methods in Applied Mechanics and Engineering*, 295, 543–561.

Bouclier, R., Elguedj, T., Combescure, A. (2015b). An isogeometric locking-free NURBS-based solid-shell element for geometrically nonlinear analysis. *International Journal for Numerical Methods in Engineering*, 101, 774–808.

Bouclier, R., Passieux, J.-C., Salaün, M. (2016). Local enrichment of NURBS patches using a non-intrusive coupling strategy: Geometric details, local refinement, inclusion, fracture. *Computer Methods in Applied Mechanics and Engineering*, 300, 1–26.

Bouclier, R., Passieux, J.-C., Salaün, M. (2017). Development of a new, more regular, mortar method for the coupling of NURBS subdomains within a NURBS patch: Application to a non-intrusive local enrichment of NURBS patches. *Computer Methods in Applied Mechanics and Engineering*, 316(1), 123–150.

Bovet, C., Parret-Fréaud, A., Spillane, N., Gosselet, P. (2017). Adaptive multipreconditioned FETI: Scalability results and robustness assessment. *Computers and Structures*, 193, 1–20.

Braibant, V. and Fleury, C. (1984). Shape optimal design using B-splines. *Computer Methods in Applied Mechanics and Engineering*, 44, 247–267.

Breitenberger, M., Apostolatos, A., Philipp, B., Wüchner, R., Bletzinger, K.-U. (2015). Analysis in computer aided design: Nonlinear isogeometric B-Rep analysis of shell structures. *Computer Methods in Applied Mechanics and Engineering*, 284, 401–457.

Brivadis, E., Buffa, A., Wohlmuth, B., Wunderlich, L. (2015). Isogeometric mortar methods. *Computer Methods in Applied Mechanics and Engineering*, 284, 292–319.

Broyden, C.G. (1970). The convergence of a class of double-rank minimization algorithms. *IMA Journal of Applied Mathematics*, 6, 76–90.

Buffa, A., Puppi, R., Vàzquez, R. (2019). A minimal stabilization procedure for isogeometric methods on trimmed geometries [Online]. Available at: https://arxiv.org/abs/1902.04937v3.

Burman, E. (2011). A penalty-free nonsymmetric Nitsche-type method for the weak imposition of boundary conditions. *SIAM Journal of Numerical Analysis*, 50, 1959–1981.

Burman, E., Claus, S., Hansbo, P., Larson, M.-G., Massing, A. (2015). CutFEM: Discretizing geometry and partial differential equations. *International Journal for Numerical Methods in Engineering*, 104, 472–501.

Calabro, F., Sangalli, G., Tani, M. (2017). Fast formation of isogeometric Galerkin matrices by weighted quadrature. *Computer Methods in Applied Mechanics and Engineering*, 316, 606–622.

Cardoso, R.P.R. and Cesar de Sa, J.M.A. (2014). Blending moving least squares techniques with NURBS basis functions for nonlinear isogeometric analysis. *Computational Mechanics*, 53(6), 1327–1340.

Caseiro, J., Valente, R., Reali, A., Kiendl, J., Auricchio, F., de Sousa, R.A. (2015). Assumed natural strain NURBS-based solid-shell element for the analysis of large deformation elasto-plastic thin-shell structures. *Computer Methods in Applied Mechanics and Engineering*, 284, 861–880.

Chantrait, T., Rannou, J., Gravouil, A. (2014). Low intrusive coupling of implicit and explicit time integration schemes for structural dynamics: Application to low energy impacts on composite structures. *Finite Elements in Analysis and Design*, 86, 23–33.

Chelaghma, S. (2018). Functionalization of Carbon/PEKK composites for aerospace applications: Characterization, modeling and influence on the composite properties. PhD thesis, Université de Toulouse.

Chen, L. and Borst, R. (2018). Adaptive refinement of hierarchical T-splines. *Computer Methods in Applied Mechanics and Engineering*, 337, 220–245.

Chevreuil, M., Nouy, A., Safatly, E. (2013). A multiscale method with patch for the solution of stochastic partial differential equations with localized uncertainties. *Computer Methods in Applied Mechanics and Engineering*, 255, 255–274.

Choi, M.-J. and Cho, S. (2015). A mesh regularization scheme to update internal control points for isogeometric shape design optimization. *Computer Methods in Applied Mechanics and Engineering*, 285, 694–713.

Choi, M.-J. and Cho, S. (2018). Constrained isogeometric design optimization of lattice structures on curved surfaces: Computation of design velocity field. *Structural and Multidisciplinary Optimization*, 58, 17–34.

Chouly, F., Fabre, M., Hild, P., Mlika, R., Pousin, J., Renard, Y. (2017). An overview of recent results on Nitsche's method for contact problems. In *Geometrically Unfitted Finite Element Methods and Applications*, Bordas, S.P.A., Burman, E., Larson, M.G., Olshanskii, M.A. (eds). Springer, Cham [Online]. Available at: https://doi.org/10.1007/978-3-319-71431-8_4.

Cirak, F., Ortiz, M., Schröder, P. (2000). Subdivision surfaces: A new paradigm for thin-shell finite-element analysis. *International Journal for Numerical Methods in Engineering*, 47(12), 2039–2072.

Cirak, F., Scott, M.J., Antonsson, E.K., Ortiz, M., Schröder, P. (2002). Integrated modeling, finite-element analysis, and engineering design for thin-shell structures using subdivision. *Computer-aided Design*, 34(2), 137–148.

Cohen, E., Lyche, T., Riesenfeld, R. (1980). Discrete B-spline and subdivision techniques in computer aided geometric design and computer graphics. *Computer Graphics and Image Processing*, 14, 87–111.

Cohen, E., Riesenfeld, R.F., Elber, G. (2001). *Geometric Modeling with Splines: An Introduction*. A.K. Peters, Natick.

Coox, L., Greco, F., Atak, O., Vandepitte, D., Desmet, W. (2017a). A robust patch coupling method for NURBS-based isogeometric analysis of non-conforming multipatch surfaces. *Computer Methods in Applied Mechanics and Engineering*, 316, 235–260.

Coox, L., Maurin, F., Greco, F., Deckers, E., Vandepitte, D., Desmet, W. (2017b). A flexible approach for coupling NURBS patches in rotationless isogeometric analysis of Kirchhoff–Love shells. *Computer Methods in Applied Mechanics and Engineering*, 325, 505–531.

Coradello, L., Loli, G., Buffa, A. (2020). A projected super-penalty method for the C^1-coupling of multi-patch isogeometric Kirchhoff plates. arXiv preprint [Online]. Available at: https://arxiv.org/abs/2007.14343.

Cormier, N., Smallwood, B., Sinclair, G., Meda, G. (1999). Aggressive submodelling of stress concentrations. *International Journal for Numerical Methods in Engineering*, 46(6), 889–909.

Cottrell, J.A., Reali, A., Bazilevs, Y., Hughes, T.J.R. (2006). Isogeometric analysis of structural vibrations. *Computer Methods in Applied Mechanics and Engineering*, 195(41–43), 5257–5296.

Cottrell, J.A., Hughes, T.J.R., Reali, A. (2007). Studies of refinement and continuity in isogeometric structural analysis. *Computer Methods in Applied Mechanics and Engineering*, 196, 4160–4183.

Cottrell, J., Hughes, T.J.R., Bazilevs, Y. (2009). *Isogeometric Analysis: Toward Integration of CAD and FEA*. Wiley, New York.

Cramer, E.J., Dennis, J.E., Frank, P.D., Lewis, R.M., Shubin, G.R. (1994). Problem formulation for multidisciplinary optimization. *SIAM Journal on Optimization*, 4(4), 754–776.

Cresta, P., Allix, O., Rey, C., Guinard, S. (2007). Nonlinear localization strategies for domain decomposition methods in structural mechanics. *Computer Methods in Applied Mechanics and Engineering*, 196, 1436–1446.

D'Angella, D., Kollmannsberger, S., Rank, E., Reali, A. (2018). Multi-level Bézier extraction for hierarchical local refinement of isogeometric analysis. *Computer Methods in Applied Mechanics and Engineering*, 328, 147–174.

Dang Manh, N., Evgrafov, A., Gravesen, J., Lahaye, D. (2014). Iso-geometric shape optimization of magnetic density separators. *COMPEL – The International Journal for Computation and Mathematics in Electrical and Electronic Engineering*, 33(4), 1416–1433.

Dauge, M., Düster, A., Rank, E. (2015). Theoretical and numerical investigation of the finite cell method. *Journal of Scientific Computing*, 65, 1039–1064.

Daxini, S.D. and Prajapati, J.M. (2017). Parametric shape optimization techniques based on meshless methods: A review. *Structural and Multidisciplinary Optimization*, 56(5), 1197–1214.

Desmorat, B. (2002). Optimisation de structures composites. PhD thesis, École normale supérieure Paris-Saclay.

Dettmer, W., Kadapa, C., Perić, D. (2016). A stabilised immersed boundary method on hierarchical B-spline grids. *Computer Methods in Applied Mechanics and Engineering*, 311, 415–437.

Dittmann, M., Schuß, S., Wohlmuth, B., Hesch, C. (2019). Weak Cn coupling for multi-patch isogeometric analysis in solid mechanics. *International Journal for Numerical Methods in Engineering*, 118(11), 678–699.

Dokken, T., Lyche, T., Pettersen, K. (2013). Polynomial splines over locally refined box-partitions. *Computer-aided Geometric Design*, 30, 331–356.

Dolean, V., Jolivet, P., Nataf, F. (2015). *An Introduction to Domain Decomposition Methods: Algorithms, Theory, and Parallel Implementation*. SIAM, Philadelphia.

Dornisch, W., Klinkel, S., Simeon, B. (2013). Isogeometric Reissner–Mindlin shell analysis with exactly calculated director vectors. *Computer Methods in Applied Mechanics and Engineering*, 253, 491–504.

Dornisch, W., Vitucci, G., Klinkel, S.O. (2015). The weak substitution method: An application of the mortar method for patch coupling in NURBS-based isogeometric analysis. *International Journal for Numerical Methods in Engineering*, 103, 205–234.

Dornisch, W., Stoeckler, J., Müller, R. (2017). Dual and approximate dual basis functions for B-splines and NURBS – Comparison and application for an efficient coupling of patches with the isogeometric mortar method. *Computer Methods in Applied Mechanics and Engineering*, 316, 449–496.

Dostál, Z., Horák, D., Kučera, R. (2006). Total FETI – An easier implementable variant of the FETI method for numerical solution of elliptic PDE. *Communications in Numerical Methods in Engineering*, 22(12), 1155–1162.

Dubois, A., Farhat, C., Abukhwejah, A.H., Shageer, H.M. (2018). Parameterization framework for the MDAO of wing structural layouts. *AIAA Journal*, 56, 1627–1638.

Duong, T.X., Roohbakhshan, F., Sauer, R.A. (2017). A new rotation-free isogeometric thin shell formulation and a corresponding continuity constraint for patch boundaries. *Computer Methods in Applied Mechanics and Engineering*, 316, 43–83.

Dureisseix, D. and Farhat, C. (2001). A numerically scalable domain decomposition method for the solution of frictionless contact problems. *International Journal for Numerical Methods in Engineering*, 50, 2643–2666.

Duval, A., Elguedj, T., Al-Akhras, H., Maurin, F. (2015). abqNURBS : implémentation d'éléments isogéométriques dans Abaqus et outils de pré- et post-traitement dédiés. *12ème Colloque national en calcul des structures*, CSMA, Giens, France, 18–22 May.

Duval, M., Passieux, J.-C., Salaün, M., Guinard, S. (2016). Non-intrusive coupling: Recent advances and scalable nonlinear domain decomposition. *Archives of Computational Methods in Engineering*, 23(1), 17–38.

Duval, M., Losinski, A., Passieux, J.-C., Salaün, M. (2018). Residual error based adaptive mesh refinement with the non-intrusive patch algorithm. *Computer Methods in Applied Mechanics and Engineering*, 329, 118–143.

Duvigneau, R. (2006). Adaptive parameterization using free-form deformation for aerodynamic shape optimization. Research report, INRIA.

Echter, R., Oesterle, B., Bischoff, M. (2013). A hierarchic family of isogeometric shell finite elements. *Computer Methods in Applied Mechanics and Engineering*, 254, 170–180.

Elber, G. (2017). Precise construction of micro-structures and porous geometry via functional composition. In *Mathematical Methods for Curves and Surfaces*, Floater, M., Lyche, T., Mazure, M.-L., Mørken, K., Schumaker, L.L. (eds). Springer International Publishing.

Elfverson, D., Larson, M., Larsson, K. (2019). A new least squares stabilized Nitsche method for cut isogeometric analysis. *Computer Methods in Applied Mechanics and Engineering*, 349, 1–16.

Elguedj, T., Bazilevs, Y., Calo, V.M., Hughes, T.J.R. (2008). B-bar and F-bar projection methods for nearly incompressible linear and non-linear elasticity and plasticity using higher-order NURBS elements. *Computer Methods in Applied Mechanics and Engineering*, 197(33–40), 2732–2762.

Evans, J.A., Bazilevs, Y., Babuška, I., Hughes, T.J.R. (2009). n-Widths, sup-infs, and optimality ratios for the k-version of the isogeometric finite element method. *Computer Methods in Applied Mechanics and Engineering*, 198(21–26), 1726–1741.

Evans, E.-J., Scott, M., Li, X., Thomas, D. (2015). Hierarchical T-splines: Analysis-suitability, Bézier extraction, and application as an adaptive basis for isogeometric analysis. *Computer Methods in Applied Mechanics and Engineering*, 284, 1–20.

Farhat, C. and Roux, F. (1991). A method of finite element tearing and interconnecting and its parallel solution algorithm. *International Journal for Numerical Methods in Engineering*, 32, 1205–1227.

Farhat, C., Mandel, J., Roux, F.X. (1994). Optimal convergence properties of the FETI domain decomposition method. *Computer Methods in Applied Mechanics and Engineering*, 115(3), 365–385.

Farhat, C., Chen, P.-S., Mandel, J., Roux, F.X. (1998). The two-level FETI method, Part II: Extension to shell problems, parallel implementation and performance results. *Computer Methods in Applied Mechanics and Engineering*, 155(1), 153–179.

Farhat, C., Lesoinne, M., Le Tallec, P., Pierson, K., Rixen, D. (2001). FETI-DP: A dual-primal unified FETI method, Part I: A faster alternative to the two-level FETI method. *International Journal for Numerical Methods in Engineering*, 50, 1523–1544.

Farin, G. (2002). *Curves and Surfaces for CAGD: A Practical Guide*. Morgan Kaufmann Publishers Inc., San Francisco.

Firl, M. (2010). Optimal shape design of shell structures. PhD thesis, Technische Universität München, Munich.

Firl, M., Wüchner, R., Bletzinger, K.-U. (2013). Regularization of shape optimization problems using FE-based parametrization. *Structural and Multidisciplinary Optimization*, 47, 507–521.

Fletcher, R. (1987). *Practical Methods of Optimization*. John Wiley & Sons, New York.

Fries, T.-P. and Omerovic, S. (2016). Higher-order accurate integration of implicit geometries. *Computer Methods in Applied Mechanics and Engineering*, 106, 323–371.

Fries, T.-P., Omerovic, S., Schöllhammer, D., Steidl, J. (2017). Higher-order meshing of implicit geometries, Part I: Integration and interpolation in cut elements. *Computer Methods in Applied Mechanics and Engineering*, 313, 759–784.

Fritz, A., Hüeber, S., Wohlmuth, B.I. (2004). A comparison of mortar and Nitsche techniques for linear elasticity. *Calcolo*, 41(3), 115–137.

Fußeder, D., Simeon, B., Vuong, A.-V. (2015). Fundamental aspects of shape optimization in the context of isogeometric analysis. *Computer Methods in Applied Mechanics and Engineering*, 286, 313–331.

Gendre, L., Allix, O., Gosselet, P. (2009). Non-intrusive and exact global/local techniques for structural problems with local plasticity. *Computational Mechanics*, 44, 233–245.

Gordon, W.J. and Riesenfeld, R.F. (1974). B-spline curves and surfaces. In *Computer-Aided Geometric Design*, Barnhill R.E. and Riesenfeld, R.F. (eds). Elsevier, New York.

Gosselet, P. and Rey, C. (2006). Non-overlapping domain decomposition methods in structural mechanics. *Archives of Computational Methods in Engineering*, 13, 515–572.

Gosselet, P., Rey, C., Pebrel, J. (2013). Total and selective reuse of Krylov subspaces for the solution to a sequence of nonlinear structural problems. *International Journal for Numerical Methods in Engineering*, 94, 60–83.

Gosselet, P., Rixen, D., Roux, F.-X., Spillane, N. (2015). Simultaneous-FETI and Block-FETI: Robust domain decomposition with multiple search directions. *International Journal for Numerical Methods in Engineering*, 104, 905–927.

Gosselet, P., Blanchard, M., Allix, O., Guguin, G. (2018). Non-invasive global-local coupling as a Schwarz domain decomposition method: Acceleration and generalization. *Advanced Modeling and Simulation in Engineering Sciences*, 5(4).

Groen, J.P., Langelaar, M., Sigmund, O., Ruess, M. (2017). Higher-order multi-resolution topology optimization using the finite cell method. *International Journal for Numerical Methods in Engineering*, 110(10), 903–920.

Guguin, G., Allix, O., Gosselet, P., Guinard, S. (2014). Nonintrusive coupling of 3D and 2D laminated composite models based on finite element 3D recovery. *International Journal for Numerical Methods in Engineering*, 98(5), 324–343.

Guguin, G., Allix, O., Gosselet, P., Guinard, S. (2016). On the computation of plate assemblies using realistic 3D joint model: A non-intrusive approach. *Advanced Modeling and Simulation in Engineering Sciences*, 3(1), 1–18.

Guinard, S., Bouclier, R., Toniolli, M., Passieux, J.-C. (2018). Multiscale analysis of complex aeronautical structures using robust non-intrusive coupling. *Advanced Modeling and Simulation in Engineering Sciences*, 5(1).

Guo, Y. (2017). Global-local model coupling for composite shell structures in the framework of isogeometric analysis. *Composite Structures*, 176, 617–629.

Guo, Y. and Ruess, M. (2015). Nitsche's method for a coupling of isogeometric thin shells and blended shell structures. *Computer Methods in Applied Mechanics and Engineering*, 284, 881–905.

Guo, Y., Ruess, M., Schillinger, D. (2017). A parameter-free variational coupling approach for trimmed isogeometric thin shells. *Computational Mechanics*, 59(4), 693–715.

Guo, Y., Heller, J., Hughes, T.J.R., Ruess, M., Schillinger, D. (2018). Variationally consistent isogeometric analysis of trimmed thin shells at finite deformations, based on the step exchange format. *Computer Methods in Applied Mechanics and Engineering*, 336, 39–79.

Gupta, P., Pereira, J., Kim, D.-J., Duarte, C., Eason, T. (2012). Analysis of three-dimensional fracture mechanics problems: A non-intrusive approach using a generalized finite element method. *Engineering Fracture Mechanics*, 90, 41–64.

Hansbo, P. (2005). Nitsche's method for interface problems in computational mechanics. *GAMM-Mitteilungen*, 28, 183–206.

Hartmann, S., Benson, D.J., Lorenz, D. (2011). About isogeometric analysis and the new NURBS-based finite elements in LS-DYNA. *8th European LS-DYNA Users Conference*, Strasbourg, France, May 23–24.

Hartmann, S., Benson, D., Nagy, A. (2016). Isogeometric analysis with LS-DYNA. *Journal of Physics: Conference Series*, 734, p. 032125.

Hasan Imam, M. (1982). Three-dimensional shape optimization. *International Journal for Numerical Methods in Engineering*, 18, 661–673.

Havens, D., Shiyekar, S., Norris, A., Bird, R.K., Kapania, R.K., Robert, O. (2011). Design, optimization, and evaluation of integrally-stiffened Al-2139 panel with curved stiffeners. Technical report 217308, NASA.

Hennig, P., Müller, S., Kästner, M. (2016). Bézier extraction and adaptive refinement of truncated hierarchical NURBS. *Computer Methods in Applied Mechanics and Engineering*, 305, 316–339.

Herrema, A.J., Wiese, N.M., Darling, C.N., Ganapathysubramanian, B., Krishnamurthy, A., Hsu, M.-C. (2017). A framework for parametric design optimization using isogeometric analysis. *Computer Methods in Applied Mechanics and Engineering*, 316, 944–965.

Herrema, A.J., Johnson, E.L., Proserpio, D.M., Wu, M.C.H., Kiendl, J., Hsu, M.-C. (2019). Penalty coupling of non-matching isogeometric Kirchhoff–Love shell patches with application to composite wind turbine blades. *Computer Methods in Applied Mechanics and Engineering*, 346, 810–840.

Hiemstra, R.R., Sangalli, G., Tani, M., Calabrò, F., Hughes, T.J.R. (2019). Fast formation and assembly of finite element matrices with application to isogeometric linear elasticity. *Computer Methods in Applied Mechanics and Engineering*, 355, 234–260.

Hirschler, T. (2019). Isogeometric modeling for the optimal design of aerostructures. PhD thesis, Université de Lyon.

Hirschler, T., Bouclier, R., Duval, A., Elguedj, T., Morlier, J. (2019a). The embedded isogeometric Kirchhoff–Love shell: From design to shape optimization of non-conforming stiffened multipatch structures. *Computer Methods in Applied Mechanics and Engineering*, 349, 774–797.

Hirschler, T., Bouclier, R., Duval, A., Elguedj, T., Morlier, J. (2019b). Isogeometric sizing and shape optimization of thin structures with a solid-shell approach. *Structural and Multidisciplinary Optimization*, 59(3), 767–785.

Hirschler, T., Bouclier, R., Dureisseix, D., Duval, A., Elguedj, T., Morlier, J. (2019c). A dual domain decomposition algorithm for the analysis of non-conforming isogeometric Kirchhoff–Love shells. *Computer Methods in Applied Mechanics and Engineering*, 357, 112578.

Hirschler, T., Bouclier, R., Duval, A., Elguedj, T., Morlier, J. (2020). A new lighting on analytical discrete sensitivities in the context of isogeometric shape optimization. *Archives of Computational Methods in Engineering*, 1–38.

Hofer, C. and Langer, U. (2017). Dual-primal isogeometric tearing and interconnecting solvers for multipatch dG-IgA equations. *Computer Methods in Applied Mechanics and Engineering*, 316, 2–21.

Horger, T., Reali, A., Wohlmuth, B., Wunderlich, L. (2019). A hybrid isogeometric approach on multi-patches with applications to Kirchhoff plates and eigenvalue problems. *Computer Methods in Applied Mechanics and Engineering*, 348, 396–408.

Hosseini, S., Remmers, J., Verhoosel, C., Borst, R. (2013). An isogeometric solid-like shell element for nonlinear analysis. *International Journal for Numerical Methods in Engineering*, 95, 238–256.

Hu, Q., Chouly, F., Hu, P., Cheng, G., Bordas, S.P. (2018). Skew-symmetric Nitsche's formulation in isogeometric analysis: Dirichlet and symmetry conditions, patch coupling and frictionless contact. *Computer Methods in Applied Mechanics and Engineering*, 341, 188–220.

Hubrich, S. and Düster, A. (2019). Numerical integration for nonlinear problems of the finite cell method using an adaptive scheme based on moment fitting. *Computers and Mathematics with Applications*, 77(7), 1983–1997.

Hubrich, S., Di Stolfo, P., Kudela, L., Kollmannsberger, S., Rank, E., Schröder, A., Düster, A. (2017). Numerical integration of discontinuous functions: Moment fitting and smart octree. *Computational Mechanics*, 60(5), 863–881.

Hughes, T.J.R., Cottrell, J., Bazilevs, Y. (2005). Isogeometric analysis: CAD, finite elements, NURBS, exact geometry and mesh refinement. *Computer Methods in Applied Mechanics and Engineering*, 194, 4135–4195.

Hughes, T.J.R., Reali, A., Sangalli, G. (2010). Efficient quadrature for NURBS-based isogeometric analysis. *Computer Methods in Applied Mechanics and Engineering*, 199(5–8), 301–313.

Ikeya, K., Shimoda, M., Shi, J.-X. (2016). Multi-objective free-form optimization for shape and thickness of shell structures with composite materials. *Composite Structures*, 135, 262–275.

Irons, B. and Tuck, R. (1969). A version of the Aitken accelerator for computer iteration. *International Journal for Numerical Methods in Engineering*, 1(3), 275–277.

Jara-Almonte, C.-C. and Knight, C.-E. (1988). The specified boundary stiffness/force SBSF method for finite element subregion analysis. *International Journal for Numerical Methods in Engineering*, 26(7), 1567–1578.

Jiang, W., Annavarapu, C., Dolbow, J., Harari, I. (2015). A robust Nitsche's formulation for interface problems with spline-based finite elements. *International Journal for Numerical Methods in Engineering*, 104, 676–696.

Jolivet, P., Hecht, F., Nataf, F., Prud'homme, C. (2014). Scalable domain decomposition preconditioners for heterogeneous elliptic problems. *Scientific Programming*, 22, 157–171.

Jomo, J., de Prenter, F., Elhaddad, M., D'Angella, D., Verhoosel, C., Kollmannsberger, S., Kirschke, J., Nübel, V., Van Brummelen, H., Rank, E. (2019). Robust and parallel scalable iterative solutions for large-scale finite cell analyses. *Finite Elements in Analysis and Design*, 163, 14–30.

Jones, E., Oliphant, T., Peterson, P. (2001). SciPy: Open source scientific tools for Python [Online]. Available at: http://www.scipy.org/.

Jongerius, S.R. and Lentink, D. (2010). Structural analysis of a dragonfly wing. *Experimental Mechanics*, 50, 1323–1334.

Joulaian, M., Hubrich, S., Düster, A. (2016). Numerical integration of discontinuities on arbitrary domains based on moment fitting. *Computational Mechanics*, 57(6), 979–999.

Jutte, C.V., Stanford, B., Wieseman, C.D., Moore, J.B. (2014). Aeroelastic tailoring of the NASA common research model via novel material and structural configurations. *52nd Aerospace Sciences Meeting*, 1–20.

Kamensky, D., Hsu, M.-C., Yu, Y., Evans, J.A., Sacks, M.S., Hughes, T.J.R. (2017). Immersogeometric cardiovascular fluid–structure interaction analysis with divergence-conforming B-splines. *Computer Methods in Applied Mechanics and Engineering*, 314, 408–472.

Kang, P. and Youn, S.-K. (2016). Isogeometric shape optimization of trimmed shell structures. *Structural and Multidisciplinary Optimization*, 53, 825–845.

Kegl, M. and Brank, B. (2006). Shape optimization of truss-stiffened shell structures with variable thickness. *Computer Methods in Applied Mechanics and Engineering*, 195(19–22), 2611–2634.

Kelley, F.-S. (1982). Mesh requirements for the analysis of a stress concentration by the specified boundary displacement method. *Proceedings of the Second International Computers in Engineering Conference*, ASME, San Diego, USA, 15–19 August.

Kelley, C.T. and Sachs, E.W. (1998). Local convergence of the symmetric rank-one iteration. *Computational Optimization and Applications*, 9, 43–63.

Kenway, G.K. and Martins, J.R. (2014). Multipoint high-fidelity aerostructural optimization of a transport aircraft configuration. *Journal of Aircraft*, 51(1), 144–160.

Kerfriden, P., Claus, S., Mihai, I. (2020). A mixed-dimensional CutFEM methodology for the simulation of fibre-reinforced composites. *Advanced Modeling and Simulation in Engineering Sciences*, 7(1), 1–26.

van Keulen, F., Haftka, R., Kim, N. (2005). Review of options for structural design sensitivity analysis, Part 1: Linear systems. *Computer Methods in Applied Mechanics and Engineering*, 194(30–33), 3213–3243.

Keye, S., Klimmek, T., Abu-Zurayk, M., Schulze, M., Ilic, C. (2017). Aero-structural optimization of the NASA common research model. *18th AIAA/ISSMO Multidisciplinary Analysis and Optimization Conference*, Denver, USA, 5–9 June.

Khalfan, H.F., Byrd, R.H., Schnabel, R.B. (1993). A theoretical and experimental study of the symmetric rank-one update. *SIAM Journal on Optimization*, 3(1), 1–24.

Kiendl, J. (2011). Isogeometric analysis and shape optimal design of shell structures. PhD thesis, Technische Universität München, Munich.

Kiendl, J., Bletzinger, K.-U., Linhard, J., Wüchner, R. (2009). Isogeometric shell analysis with Kirchhoff–Love elements. *Computer Methods in Applied Mechanics and Engineering*, 198(49), 3902–3914.

Kiendl, J., Bazilevs, Y., Hsu, M.-C., Wüchner, R., Bletzinger, K.-U. (2010). The bending strip method for isogeometric analysis of Kirchhoff–Love shell structures comprised of multiple patches. *Computer Methods in Applied Mechanics and Engineering*, 199(37), 2403–2416.

Kiendl, J., Schmidt, R., Wüchner, R., Bletzinger, K.-U. (2014). Isogeometric shape optimization of shells using semi-analytical sensitivity analysis and sensitivity weighting. *Computer Methods in Applied Mechanics and Engineering*, 274, 148–167.

Kiendl, J., Hsu, M.-C., Wu, M.C., Reali, A. (2015). Isogeometric Kirchhoff–Love shell formulations for general hyperelastic materials. *Computer Methods in Applied Mechanics and Engineering*, 291, 280–303.

Kikis, G., Dornisch, W., Klinkel, S. (2019). Adjusted approximation spaces for the treatment of transverse shear locking in isogeometric Reissner–Mindlin shell analysis. *Computer Methods in Applied Mechanics and Engineering*, 354, 850–870.

Kirchhoff, G. (1850). Über das Gleichgewicht und die Bewegung einer elastischen Scheibe. *Journal für die reine und angewandte Mathematik*, 40, 51–88.

Klawonn, A. and Widlund, O. (2001). FETI and Neumann–Neumann iterative substructuring methods: Connections and new results. *Communications on Pure and Applied Mathematics*, 54(1), 57–90.

Klawonn, A., Lanser, M., Rheinbach, O. (2015). Toward extremely scalable nonlinear domain decomposition methods for elliptic partial differential equations. *SIAM Journal on Scientific Computing*, 37(6), C667–C696.

Kleiss, S.K., Pechstein, C., Jüttler, B., Tomar, S. (2012). IETI – Isogeometric tearing and interconnecting. *Computer Methods in Applied Mechanics and Engineering*, 247–248, 201–215.

Kollmannsberger, S., Özcan, A., Baiges, J., Ruess, M., Rank, E., Reali, A. (2015). Parameter-free, weak imposition of Dirichlet boundary conditions and coupling of trimmed and non-conforming patches. *International Journal for Numerical Methods in Engineering*, 101, 670–699.

Kozubek, T., Vondrák, V., Menšík, M., Horák, D., Dostál, Z., Hapla, V., Kabelíková, P., Čermák, M. (2013). Total FETI domain decomposition method and its massively parallel implementation. *Advances in Engineering Software*, 60–61, 14–22.

Kraft, D. (1988). A software package for sequential quadratic programming. *Deutsche Forschungs- und Versuchsanstalt für Luft- und Raumfahrt*, 88(28).

Kudela, L., Zander, N., Bog, T., Kollmannsberger, S., Rank, E. (2015). Efficient and accurate numerical quadrature for immersed boundary method. *Advanced Modeling and Simulation in Engineering Sciences*, 2, 110.

Küttler, U. and Wall, W. (2008). Fixed-point fluid–structure interaction solvers with dynamic relaxation. *Computational Mechanics*, 43(1), 61–72.

Lacour, C. (1997). Analyse et résolution numérique de méthodes de sous-domaines non conformes pour des problèmes de plaques. PhD thesis, Université Pierre et Marie Curie – Paris VI.

Ladeveze, P., Arnaud, L., Rouch, P., Blanzé, C. (2001). The variational theory of complex rays for the calculation of medium-frequency vibrations. *Engineering Computations*, 18, 193–214.

Lagarias, J.C., Reeds, J.A., Wright, M.H., Wright, P.E. (1998). Convergence properties of the Nelder–Mead simplex method in low dimensions. *SIAM Journal on Optimization*, 9(1), 112–147.

Lai, Y., Zhang, Y.J., Liu, L., Wei, X., Fang, E., Lua, J. (2017). Integrating CAD with Abaqus: A practical isogeometric analysis software platform for industrial applications. *Computers & Mathematics with Applications*, 74(7), 1648–1660.

Lassila, T. and Rozza, G. (2010). Parametric free-form shape design with PDE models and reduced basis method. *Computer Methods in Applied Mechanics and Engineering*, 199(23–24), 1583–1592.

Le, C., Bruns, T., Tortorelli, D. (2011). A gradient-based, parameter-free approach to shape optimization. *Computer Methods in Applied Mechanics and Engineering*, 200(9–12), 985–996.

Lee, B.-G. and Park, Y. (2002). Degree elevation of NURBS curves by weighted blossom. *Korean Journal of Computational & Applied Mathematics*, 9(1), 151–165.

Legrain, G. (2013). A NURBS enhanced extended finite element approach for unfitted CAD analysis. *Computational Mechanics*, 52, 913–929.

Lei, Z., Gillot, F., Jezequel, L. (2018). Shape optimization for natural frequency with isogeometric Kirchhoff–Love shell and sensitivity mapping. *Mathematical Problems in Engineering*, 2018, 1–11.

Leidinger, L., Breitenberger, M., Bauer, A., Hartmann, S., Wüchner, R., Bletzinger, K.-U., Duddeck, F., Song, L. (2019). Explicit dynamic isogeometric B-Rep analysis of penalty-coupled trimmed NURBS shells. *Computer Methods in Applied Mechanics and Engineering*, 351, 891–927.

Leonetti, L., Liguori, F.S., Magisano, D., Kiendl, J., Reali, A., Garcea, G. (2020). A robust penalty coupling of non-matching isogeometric Kirchhoff–Love shell patches in large deformations. *Computer Methods in Applied Mechanics and Engineering*, 371, 113289.

Le Tallec, P., De Roeck, Y.-H., Vidrascu, M. (1991). Domain-decomposition methods for large linearly elliptic three dimensional problems. *Journal of Computational and Applied Mathematics*, 34, 93–117.

Lew, A.J. and Buscaglia, G.C. (2008). A discontinuous-Galerkin-based immersed boundary method. *International Journal for Numerical Methods in Engineering*, 76(4), 427–454.

Li, N., Link, G., Wang, T., Ramopoulos, V., Neumaier, D., Hofele, J., Walter, M., Jelonnek, J. (2020). Path-designed 3D printing for topological optimized continuous carbon fibre reinforced composite structures. *Composites Part B: Engineering*, 182, 107612.

Lian, H., Kerfriden, P., Bordas, S. (2017). Shape optimization directly from CAD: An isogeometric boundary element approach using T-splines. *Computer Methods in Applied Mechanics and Engineering*, 317, 1–41.

Lipton, S., Evans, J.A., Bazilevs, Y., Elguedj, T., Hughes, T.J.R. (2010). Robustness of isogeometric structural discretizations under severe mesh distortion. *Computer Methods in Applied Mechanics and Engineering*, 199(5–8), 357–373.

Liu, Y.J., Sun, Q., Fan, X.L. (2014). A non-intrusive global/local algorithm with non-matching interface: Derivation and numerical validation. *Computer Methods in Applied Mechanics and Engineering*, 277, 81–103.

Liu, X., Réthoré, J., Baietto, M.-C., Sainsot, P., Lubrecht, A.A. (2019). An efficient strategy for large scale 3D simulation of heterogeneous materials to predict effective thermal conductivity. *Computational Materials Science*, 166, 265–275.

Locatelli, D., Mulani, S.B., Kapania, R.K. (2014). Parameterization of curvilinear spars and ribs for optimum wing structural design. *Journal of Aircraft*, 51(2), 532–546.

Love, A.E.H. (1888). The small free vibrations and deformation of a thin elastic shell. *Philosophical Transactions of the Royal Society of London (A)*, 179, 491–546.

Lund, E. and Stegmann, J. (2006). Eigenfrequency and buckling optimization of laminated composite shell structures using discrete material optimization. In *IUTAM Symposium on Topological Design Optimization of Structures, Machines and Materials: Status and Perspectives*, Bendsøe, M.P., Olhoff, N., Sigmund, O. (eds). Springer, New York.

Lyu, Z. and Martins, J.R. (2015). Aerodynamic shape optimization of an adaptive morphing trailing-edge wing. *Journal of Aircraft*, 52(6), 1951–1970.

Mandel, J. (1993). Balancing domain decomposition. *Communications in Numerical Methods in Engineering*, 9, 233–241.

Maquart, T. (2019). Trivariate models generation from unstructured surface manifolds for isogeometric analysis – Application to reduced order modeling with geometric parameters. PhD thesis, Université de Lyon.

Martins, J.R. and Lambe, A.B. (2013). Multidisciplinary design optimization: A survey of architectures. *AIAA Journal*, 51(9), 2049–2075.

Marussig, B. and Hughes, T.J.R. (2018). A review of trimming in isogeometric analysis: Challenges, data exchange and simulation aspects. *Archives of Computational Methods in Engineering*, 25, 1059–1127.

Massarwi, F. and Elber, G. (2016). A B-spline based framework for volumetric object modeling. *Computer-aided Design*, 78, 36–47.

Massarwi, F., Machchhar, J., Antolin, P., Elber, G. (2018). Hierarchical, random and bifurcation tiling with heterogeneity in micro-structures construction via functional composition. *Computer-aided Design*, 102, 148–159.

Massarwi, F., Antolin, P., Elber, G. (2019). Volumetric untrimming: Precise decomposition of trimmed trivariates into tensor products. *Computer-aided Geometric Design*, 71, 1–15.

Maurin, F., Greco, F., Coox, L., Vandepitte, D., Desmet, W. (2018). Isogeometric collocation for Kirchhoff–Love plates and shells. *Computer Methods in Applied Mechanics and Engineering*, 329, 396–420.

McNeel, R. & Associates (n.d.). Rhinoceros 3D [Online]. Available at: https://www.rhino3d.com/ [Accessed 31 March 2021].

Morganti, S., Auricchio, F., Benson, D., Gambarin, F., Hartmann, S., Hughes, T.J.R., Reali, A. (2015). Patient-specific isogeometric structural analysis of aortic valve closure. *Computer Methods in Applied Mechanics and Engineering*, 284, 508–520.

Mulani, S., Slemp, W., Kapania, R. (2013). EBF3PanelOpt: An optimization framework for curvilinear blade-stiffened panels. *Thin-walled Structures*, 63, 13–26.

Müller, B., Kummer, F., Oberlack, M. (2013). Highly accurate surface and volume integration on implicit domains by means of moment-fitting. *International Journal for Numerical Methods in Engineering*, 96, 512–528.

Nagy, A.P. (2011). Isogeometric design optimisation. PhD thesis, Technische Universiteit Delft.

Nagy, A.P. and Benson, D.-J. (2015). On the numerical integration of trimmed isogeometric elements. *Computer Methods in Applied Mechanics and Engineering*, 284, 165–185.

Nagy, A.P., Abdalla, M.M., Gürdal, Z. (2010). Isogeometric sizing and shape optimisation of beam structures. *Computer Methods in Applied Mechanics and Engineering*, 199(17–20), 1216–1230.

Nagy, A.P., Abdalla, M.M., Gürdal, Z. (2011). Isogeometric design of elastic arches for maximum fundamental frequency. *Structural and Multidisciplinary Optimization*, 43(1), 135–149.

Nagy, A.P., Ijsselmuiden, S.T., Abdalla, M.M. (2013). Isogeometric design of anisotropic shells: Optimal form and material distribution. *Computer Methods in Applied Mechanics and Engineering*, 264, 145–162.

Naylor, R. (2019). Prédiction des premiers endommagements des composites tissés par modélisation à l'échelle microscopique et validation multi-échelle. PhD thesis, Université Paris–Saclay.

Néron, D., Boucard, P.-A., Relun, N. (2015). Time-space PGD for the rapid solution of 3D nonlinear parametrized problems in the many-query context. *International Journal for Numerical Methods in Engineering*, 103, 275–292.

Nguyen, D.M., Evgrafov, A., Gravesen, J. (2012). Isogeometric shape optimization for electromagnetic scattering problems. *Progress in Electromagnetics Research B*, 45, 117–146.

Nguyen, V.-P., Kerfriden, P., Brino, M., Bordas, S.P.A., Bonisoli, E. (2014). Nitsche's method for two and three dimensional NURBS patch coupling. *Computational Mechanics*, 53, 1163–1182.

Nguyen-Thanh, N., Zhou, K., Zhuang, X., Areias, P., Nguyen-Xuan, H., Bazilevs, Y., Rabczuk, T. (2017). Isogeometric analysis of large-deformation thin shells using RHT-splines for multiple-patch coupling. *Computer Methods in Applied Mechanics and Engineering*, 316, 1157–1178.

Nocedal, J. and Wright, S.J. (2006). *Numerical Optimization*, 2nd edition. Springer, New York.

Nouy, A. and Pled, F. (2018). A multiscale method for semi-linear elliptic equations with localized uncertainties and non-linearities. *ESAIM: Mathematical Modelling and Numerical Analysis – Modélisation mathématique et analyse numérique*, 52(5), 1763–1802.

Oden, J.T., Babuska, I., Baumann, C.E. (1998). A discontinuous hp finite element method for diffusion problems. *Journal of Computational Physics*, 146, 491–519.

Olhoff, N., Bendsøe, M., Rasmussen, J. (1991). On CAD-integrated structural topology and design optimization. *Computer Methods in Applied Mechanics and Engineering*, 89, 259–279.

Olhoff, N., Rasmussen, J., Lund, E. (1993). A method of "exact" numerical differentiation for error elimination in finite-element-based semi-analytical shape sensitivity analyses*. *Mechanics of Structures and Machines*, 21(1), 1–66.

Oumaziz, P., Gosselet, P., Boucard, P.-A., Abbas, M. (2018). A parallel non-invasive multiscale strategy for a mixed domain decomposition method with frictional contact. *International Journal for Numerical Methods in Engineering*, 115(8), 893–912.

Papadrakakis, M., Lagaros, N.D., Tsompanakis, Y., Plevris, V. (2001). Large scale structural optimization: Computational methods and optimization algorithms. *Archives of Computational Methods in Engineering*, 8(3), 239–301.

Papapetrou, V.S., Patel, C., Tamijani, A.Y. (2020). Stiffness-based optimization framework for the topology and fiber paths of continuous fiber composites. *Composites Part B: Engineering*, 183, 107681.

Park, B.-U., Seo, Y.-D., Sigmund, O., Youn, S.-K. (2013). Shape optimization of the stokes flow problem based on isogeometric analysis. *Structural and Multidisciplinary Optimization*, 48(5), 965–977.

Parvizian, J., Düster, A., Rank, E. (2012). Topology optimization using the finite cell method. *Optimization and Engineering*, 13(1), 57–78.

Passieux, J.-C., Réthoré, J., Gravouil, A., Baietto, M.-C. (2013). Local/global non-intrusive crack propagation simulation using a multigrid X-FEM solver. *Computational Mechanics*, 56(2), 1381–1393.

Passieux, J.-C., Bugarin, F., David, C., Périé, J.-N., Robert, L. (2015). Multiscale displacement field measurement using digital image correlation: Application to the identification of elastic properties. *Experimental Mechanics*, 55(1), 121–137.

Patelli, A.S., Dedè, L., Lassila, T., Bartezzaghi, A., Quarteroni, A. (2017). Isogeometric approximation of cardiac electrophysiology models on surfaces: An accuracy study with application to the human left atrium. *Computer Methods in Applied Mechanics and Engineering*, 317, 248–273.

Paul, K., Zimmermann, C., Duong, T.X., Sauer, R.A. (2020). Isogeometric continuity constraints for multi-patch shells governed by fourth-order deformation and phase field models. *Computer Methods in Applied Mechanics and Engineering*, 370, 113219

Piegl, L.A. (2005). Ten challenges in computer-aided design. *Computer-aided Design*, 37(4), 461–470.

Piegl, L.A. and Tiller, W. (1997). *The NURBS Book*, 2nd edition. Springer, New York.

de Prenter, F., Verhoosel, C.-V., van Zwieten, G.-J., van Brummelen, V.-H. (2017). Condition number analysis and preconditioning of the finite cell method. *Computer Methods in Applied Mechanics and Engineering*, 316, 297–327.

Qian, X. (2010). Full analytical sensitivities in NURBS based isogeometric shape optimization. *Computer Methods in Applied Mechanics and Engineering*, 199, 2059–2071.

Ramm, E., Bletzinger, K.-U., Reitinger, R. (1993). Shape optimization of shell structures. *Revue européenne des éléments finis*, 2(3), 377–398.

Rank, E., Ruess, M., Kollmannsberger, S., Schillinger, D., Düster, A. (2012). Geometric modeling, isogeometric analysis and the finite cell method. *Computer Methods in Applied Mechanics and Engineering*, 249, 104–115.

Renard, M. (2018). STELIA Aerospace imprime en 3D un panneau de fuselage [Online]. Available at: https://www.3dnatives.com/stelia-aerospace-21022018/.

Réthoré, J. (2015). Automatic crack tip detection and stress intensity factors estimation of curved cracks from digital images. *International Journal for Numerical Methods in Engineering*, 103(7), 516–534.

Rivière, B. (2008). *Discontinuous Galerkin Methods For Solving Elliptic and Parabolic Equations: Theory and Implementation*. SIAM, Philadelphia.

Rivière, B., Wheeler, M.F., Girault, V. (2001). *A priori* error estimates for finite element methods based on discontinuous approximation spaces for elliptic problems. *SIAM Journal on Numerical Analysis*, 39, 902–931.

Rixen, D.J. (2002). Extended preconditioners for the FETI method applied to constrained problems. *International Journal for Numerical Methods in Engineering*, 54(1), 1–26.

Rixen, D.J. and Farhat, C. (1999). A simple and efficient extension of a class of substructure based preconditioners to heterogeneous structural mechanics problems. *International Journal for Numerical Methods in Engineering*, 44(4), 489–516.

Rixen, D.J., Farhat, C., Tezaur, R., Mandel, J. (1999). Theoretical comparison of the FETI and algebraically partitioned FETI methods, and performance comparisons with a direct sparse solver. *International Journal for Numerical Methods in Engineering*, 46(4), 501–533.

Rogers, D.F. (2000). *An Introduction to NURBS: With Historical Perspective*. Elsevier, San Diego, CA.

Ruess, M., Schillinger, D., Özcan, A.-I., Rank, E. (2014). Weak coupling for isogeometric analysis of non-matching and trimmed multi-patch geometries. *Computer Methods in Applied Mechanics and Engineering*, 169, 46–71.

Saad, M.-H. (2009). *Elasticity, Theory, Applications, and Numerics*. Academic Press, Oxford.

Sanders, J., Laursen, T., Puso, M. (2012). A Nitsche embedded mesh method. *Computational Mechanics*, 49, 243–257.

Sangalli, G. and Tani, M. (2018). Matrix-free weighted quadrature for a computationally efficient isogeometric k-method. *Computer Methods in Applied Mechanics and Engineering*, 338, 117–133.

Savenkov, E.B., Andrae, H., Iliev, O.P., Iliev, H. (2008). An analysis of one regularization approach for solution of pure Neumann problem. Technical report, Fraunhofer-Institut für Techno- und Wirtschaftsmathematik.

Schillinger, D. and Ruess, M. (2015). The finite cell method: A review in the context of higher-order structural analysis of CAD and image-based geometric models. *Archives of Computational Methods in Engineering*, 22(3), 391–455.

Schillinger, D., Hossain, S.J., Hughes, T.J.R. (2014). Reduced Bézier element quadrature rules for quadratic and cubic splines in isogeometric analysis. *Computer Methods in Applied Mechanics and Engineering*, 277, 1–45.

Schillinger, D., Harari, I., Hsu, M.-C., Kamensky, D., Stoter, K.F.S., Yu, Y., Ying, Z. (2016a). The non-symmetric Nitsche method for the parameter-free imposition of weak boundary and coupling conditions in immersed finite elements. *Computer Methods in Applied Mechanics and Engineering*, 309, 625–652.

Schillinger, D., Ruthala, P.K., Nguyen, L.H. (2016b). Lagrange extraction and projection for NURBS basis functions: A direct link between isogeometric and standard nodal finite element formulations. *International Journal for Numerical Methods in Engineering*, 108(6), 515–534.

Schmidt, R. (2013). Trimming, mapping, and optimization in isogeometric analysis of shell structures. PhD thesis, Technische Universität München.

Schöllhammer, D. and Fries, T.P. (2019). Kirchhoff–Love shell theory based on tangential differential calculus. *Computational Mechanics*, 64(1), 113–131.

Schöllhammer, D., Marussig, B., Fries, T.P. (2019). An isogeometric fictitious domain method for trimmed Kirchhoff–Love shells using extended B-splines and Nitsche's method. *7th International Conference on Isogeometric Analysis*, Munich, Germany, 18–20 September.

Schuß, S., Dittmann, M., Wohlmuth, B., Klinkel, S., Hesch, C. (2019). Multi-patch isogeometric analysis for Kirchhoff–Love shell elements. *Computer Methods in Applied Mechanics and Engineering*, 349, 91–116.

Scott, M., Li, X., Sederberg, T., Hughes, T.J.R. (2012). Local refinement of analysis-suitable T-splines. *Computer Methods in Applied Mechanics and Engineering*, 213–216, 206–222.

Scott, M., Thomas, D., Evans, E.-J. (2014). Isogeometric spline forests. *Computer Methods in Applied Mechanics and Engineering*, 269, 222–264.

Sederberg, T.W. and Parry, S.R. (1986). Free-form deformation of solid geometric models. *ACM SIGGRAPH Computer Graphics*, 20(4), 151–160.

Seitz, A., Farah, P., Kremheller, J., Wohlmuth, B.I., Wall, W.A., Popp, A. (2016). Isogeometric dual mortar methods for computational contact mechanics, *Computer Methods in Applied Mechanics and Engineering*, 301, 259–280.

Seo, Y.-D., Kim, H.-J., Youn, S.-K. (2010). Isogeometric topology optimization using trimmed spline surfaces. *Computer Methods in Applied Mechanics and Engineering*, 199(49–52), 3270–3296.

Shojaee, S., Izadpanah, E., Valizadeh, N., Kiendl, J. (2012). Free vibration analysis of thin plates by using a NURBS-based isogeometric approach. *Finite Elements in Analysis and Design*, 61, 23–34.

Shroff, S., Acar, E., Kassapoglou, C. (2017). Design, analysis, fabrication, and testing of composite grid-stiffened panels for aircraft structures. *Thin-walled Structures*, 119, 235–246.

Singh, K., Zhao, W., Kapania, R.K. (2017). An optimization framework for curvilinearly stiffened composite pressure vessels and pipes. *ASME 2017 Pressure Vessels and Piping Conference*, Waikoloa, USA, 16–20 July.

Sommerwerk, K., Woidt, M., Haupt, M.-C., Horst, P. (2017). Reissner–Mindlin shell implementation and energy conserving isogeometric multi-patch coupling. *International Journal for Numerical Methods in Engineering*, 109, 982–1012.

van Sosin, B. and Elber, G. (2018). Crossing knot lines in composition of freeform B-spline geometry. *Computer-aided Geometric Design*, 62, 217–227.

Spillane, N. and Rixen, D. (2013). Automatic spectral coarse spaces for robust finite element tearing and interconnecting and balanced domain decomposition algorithms. *International Journal for Numerical Methods in Engineering*, 95(11), 953–990.

Spillane, N., Dolean, V., Hauret, P., Nataf, F., Rixen, D.J. (2013). Solving generalized eigenvalue problems on the interfaces to build a robust two-level FETI method. *Comptes rendus mathématique*, 351(5–6), 197–201.

Srinivasan, S., Biggers, S.-B., Latour, R.-A. (1996). Identifying global/local interface boundaries using an objective search method. *International Journal for Numerical Methods in Engineering*, 39, 805–828.

Stanford, B., Beran, P., Bhatia, M. (2014). Aeroelastic topology optimization of blade-stiffened panels. *Journal of Aircraft*, 51(3), 938–944.

Stavrev, A., Nguyen, L.-H., Shen, R., Varduhn, V., Behr, M., Elgeti, S., Schillinger, D. (2016). Geometrically accurate, efficient, and flexible quadrature techniques for the tetrahedral finite cell method. *Computer Methods in Applied Mechanics and Engineering*, 310, 646–673.

Stavroulakis, G., Tsapetis, D., Papadrakakis, M. (2018). Non-overlapping domain decomposition solution schemes for structural mechanics isogeometric analysis. *Computer Methods in Applied Mechanics and Engineering*, 341, 695–717.

Stefanica, D. (2001). A numerical study of FETI algorithms for mortar finite element methods. *SIAM Journal on Scientific Computing*, 23(4), 1135–1160.

Stefanica, D. (2005). Parallel FETI algorithms for mortars. *Applied Numerical Mathematics*, 54(2), 266–279.

Stroud, I. (2006). *Boundary Representation Modelling Techniques*. Springer Science & Business Media, Lausanne.

Taheri, A. and Hassani, B. (2014). Simultaneous isogeometrical shape and material design of functionally graded structures for optimal eigenfrequencies. *Computer Methods in Applied Mechanics and Engineering*, 277, 46–80.

Teschemacher, T., Bauer, A.M., Oberbichler, T., Breitenberger, M., Rossi, R., Wüchner, R., Bletzinger, K.-U. (2018). Realization of CAD-integrated shell simulation based on isogeometric B-rep analysis. *Advanced Modeling and Simulation in Engineering Sciences*, 5(1), 19.

Thomas, D., Engvall, L., Schmidt, S., Tew, K., Scott, M. (2018). U-splines: Splines over unstructured meshes. Preprint [Online]. Available at: https://coreform.com/papers/usplines.pdf

Tirvaudey, M. (2019). Couplage AIG/MEF pour l'analyse de détails structuraux par une approche non-intrusive et certifiée. PhD thesis, Université de Toulouse.

Tirvaudey, M., Chamoin, L., Bouclier, R., Passieux, J.-C. (2020). *A posteriori* error estimation and adaptivity in non-intrusive couplings between concurrent models. *Computer Methods in Applied Mechanics and Engineering*, 367, 113104.

Toselli, A. and Widlund, O. (2006). *Domain Decomposition Methods – Algorithms and Theory*, Springer Science & Business Media, Zurich.

Totaro, G. and Nicola, F.D. (2012). Recent advance on design and manufacturing of composite anisogrid structures for space launchers. *Acta Astronautica*, 81(2), 570–577.

Vassberg, J., Dehaan, M., Rivers, M., Wahls, R. (2008). Development of a common research model for applied CFD validation studies. *26th AIAA Applied Aerodynamics Conference*, Honolulu, USA, 18–21 August.

Veiga, L., Buffa, A., Cho, D., Sangalli, G. (2012). Analysis-suitable T-splines are dual-compatible. *Computer Methods in Applied Mechanics and Engineering*, 249, 42–51.

Verhoosel, C.-V., van Zwieten, G.-J., van Rietbergen, B., de Borst, R. (2015). Image-based goal-oriented adaptive isogeometric analysis with application to the micro-mechanical modeling of trabecular bone. *Computer Methods in Applied Mechanics and Engineering*, 284, 138–164.

Vilanova, G., Colominas, I., Gomez, H. (2013). Capillary networks in tumor angiogenesis: From discrete endothelial cells to phase-field averaged descriptions via isogeometric analysis. *International Journal for Numerical Methods in Biomedical Engineering*, 29(10), 1015–1037.

Vuong, A.-V., Giannelli, C., Jüttler, B., Simeon, B. (2011). A hierarchical approach to adaptive local refinement in isogeometric analysis. *Computer Methods in Applied Mechanics and Engineering*, 200, 3554–3567.

Wall, W.A., Frenzel, M.A., Cyron, C. (2008). Isogeometric structural shape optimization. *Computer Methods in Applied Mechanics and Engineering*, 197(33), 2976–2988.

Wang, Z.-P. and Turteltaub, S. (2015). Isogeometric shape optimization for quasi-static processes. *International Journal for Numerical Methods in Engineering*, 104, 347–371.

Wang, W., Clausen, P.M., Bletzinger, K.-U. (2015). Improved semi-analytical sensitivity analysis using a secant stiffness matrix for geometric nonlinear shape optimization. *Computers & Structures*, 146, 143–151.

Wang, Z.P., Poh, L.H., Dirrenberger, J., Zhu, Y., Forest, S. (2017a). Isogeometric shape optimization of smoothed petal auxetic structures via computational periodic homogenization. *Computer Methods in Applied Mechanics and Engineering*, 323, 250–271.

Wang, Z.-P., Turteltaub, S., Abdalla, M. (2017b). Shape optimization and optimal control for transient heat conduction problems using an isogeometric approach. *Computers & Structures*, 185, 59–74.

Wang, Y., Wang, Z., Xia, Z., Poh, L.H. (2018). Structural design optimization using isogeometric analysis: A comprehensive review. *Computer Modeling in Engineering and Sciences*, 117, 455–507.

Wang, D., Abdalla, M.M., Wang, Z.-P., Su, Z. (2019). Streamline stiffener path optimization (SSPO) for embedded stiffener layout design of non-uniform curved grid-stiffened composite (NCGC) structures. *Computer Methods in Applied Mechanics and Engineering*, 344, 1021–1050.

Weeger, O., Narayanan, B., Dunn, M.L. (2019). Isogeometric shape optimization of nonlinear, curved 3D beams and beam structures. *Computer Methods in Applied Mechanics and Engineering*, 345, 26–51.

Whitcomb, J. (1991). Iterative global/local finite element analysis. *Computers and Structures*, 40(4), 1027–1031.

Wohlmuth, B.I. (2000). A mortar finite element method using dual spaces for the Lagrange multiplier. *SIAM Journal on Numerical Analysis*, 38(3), 989–1012.

Wohlmuth, B.I. (2001). *Discretization Methods and Iterative Solvers Based on Domain Decomposition*. Springer Berlin Heidelberg, Berlin/Heidelberg.

Wunderlich, L., Seitz, A., Alaydin, M., Wohlmuth, B., Popp, A. (2019). Biorthogonal splines for optimal weak patch-coupling in isogeometric analysis with applications to finite deformation elasticity. *Computer Methods in Applied Mechanics and Engineering*, 346, 197–215.

Xu, G., Mourrain, B., Duvigneau, R., Galligo, A. (2013). Analysis-suitable volume parameterization of multi-block computational domain in isogeometric applications. *Computer-aided Design*, 45, 395–404.

Xu, J., Sun, N., Shu, L., Rabczuk, T., Xu, G. (2017). An improved isogeometric analysis method for trimmed geometries. arXiv preprint [Online]. Available at: https://arxiv.org/abs/1707.00323.

Yera, R., Rossi, N., Méndez, C., Huespe, A.E. (2020). Topology design of 2D and 3D elastic material microarchitectures with crystal symmetries displaying isotropic properties close to their theoretical limits. *Applied Materials Today*, 18, 100456.

Zhao, W. and Kapania, R.K. (2016). Buckling analysis of unitized curvilinearly stiffened composite panels. *Composite Structures*, 135, 365–382, Amsterdam.

Zienkiewicz, O.C., Taylor, R.L., Zhu, J.Z. (2005). *The Finite Element Method: Its Basis and Fundamentals*. Elsevier.

Zou, Z., Scott, M., Borden, M.J., Thomas, D.C., Dornisch, W., Brivadis, E. (2018). Isogeometric Bézier dual mortaring: Refineable higher-order spline dual bases and weakly continuous geometry. *Computer Methods in Applied Mechanics and Engineering*, 333, 497–534.

Index

P

S, T

Other titles from

iSTE

in

Numerical Methods in Engineering

2022

GRANGE Stéphane, SALCIARINI Diana
Deterministic Numerical Modeling of Soil–Structure Interaction

2021

GENTIL Christian, GOUATY Gilles, SOKOLOV Dmitry
Geometric Modeling of Fractal Forms for CAD
(Geometric Modeling and Applications Set – Volume 5)

2020

GEORGE Paul Louis, ALAUZET Frédéric, LOSEILLE Adrien, MARÉCHAL Loïc
Meshing, Geometric Modeling and Numerical Simulation 3: Storage, Visualization and In Memory Strategies
(Geometric Modeling and Applications Set – Volume 4)

SIGRIST Jean-François
Numerical Simulation, An Art of Prediction 2: Examples

2019

DA Daicong
Topology Optimization Design of Heterogeneous Materials and Structures

GEORGE Paul Louis, BOROUCHAKI Houman, ALAUZET Frédéric, LAUG Patrick, LOSEILLE Adrien, MARÉCHAL Loïc
Meshing, Geometric Modeling and Numerical Simulation 2: Metrics, Meshes and Mesh Adaptation
(Geometric Modeling and Applications Set – Volume 2)

MARI Jean-Luc, HÉTROY-WHEELER Franck, SUBSOL Gérard
Geometric and Topological Mesh Feature Extraction for 3D Shape Analysis
(Geometric Modeling and Applications Set – Volume 3)

SIGRIST Jean-François
Numerical Simulation, An Art of Prediction 1: Theory

2017

BOROUCHAKI Houman, GEORGE Paul Louis
Meshing, Geometric Modeling and Numerical Simulation 1: Form Functions, Triangulations and Geometric Modeling
(Geometric Modeling and Applications Set – Volume 1)

2016

KERN Michel
Numerical Methods for Inverse Problems

ZHANG Weihong, WAN Min
Milling Simulation: Metal Milling Mechanics, Dynamics and Clamping Principles

2015

ANDRÉ Damien, CHARLES Jean-Luc, IORDANOFF Ivan
3D Discrete Element Workbench for Highly Dynamic Thermo-mechanical Analysis
(Discrete Element Model and Simulation of Continuous Materials Behavior Set – Volume 3)

JEBAHI Mohamed, ANDRÉ Damien, TERREROS Inigo, IORDANOFF Ivan
Discrete Element Method to Model 3D Continuous Materials (Discrete Element Model and Simulation of Continuous Materials Behavior Set – Volume 1)

JEBAHI Mohamed, DAU Frédéric, CHARLES Jean-Luc, IORDANOFF Ivan
Discrete-continuum Coupling Method to Simulate Highly Dynamic Multi-scale Problems: Simulation of Laser-induced Damage in Silica Glass (Discrete Element Model and Simulation of Continuous Materials Behavior Set – Volume 2)

SOUZA DE CURSI Eduardo
Variational Methods for Engineers with Matlab®

2014

BECKERS Benoit, BECKERS Pierre
Reconciliation of Geometry and Perception in Radiation Physics

BERGHEAU Jean-Michel
Thermomechanical Industrial Processes: Modeling and Numerical Simulation

BONNEAU Dominique, FATU Aurelian, SOUCHET Dominique
Hydrodynamic Bearings – Volume 1
Mixed Lubrication in Hydrodynamic Bearings – Volume 2
Thermo-hydrodynamic Lubrication in Hydrodynamic Bearings – Volume 3
Internal Combustion Engine Bearings Lubrication in Hydrodynamic Bearings – Volume 4

DESCAMPS Benoît
Computational Design of Lightweight Structures: Form Finding and Optimization

2013

YASTREBOV Vladislav A.
Numerical Methods in Contact Mechanics

2012

DHATT Gouri, LEFRANÇOIS Emmanuel, TOUZOT Gilbert
Finite Element Method

SAGUET Pierre
Numerical Analysis in Electromagnetics

SAANOUNI Khemais
Damage Mechanics in Metal Forming: Advanced Modeling and Numerical Simulation

2011

CHINESTA Francisco, CESCOTTO Serge, CUETO Elias, LORONG Philippe
Natural Element Method for the Simulation of Structures and Processes

DAVIM Paulo J.
Finite Element Method in Manufacturing Processes

POMMIER Sylvie, GRAVOUIL Anthony, MOËS Nicolas, COMBESCURE Alain
Extended Finite Element Method for Crack Propagation

2010

SOUZA DE CURSI Eduardo, SAMPAIO Rubens
Modeling and Convexity

2008

BERGHEAU Jean-Michel, FORTUNIER Roland
Finite Element Simulation of Heat Transfer

EYMARD Robert
Finite Volumes for Complex Applications V: Problems and Perspectives

FREY Pascal, GEORGE Paul Louis
Mesh Generation: Application to finite elements – 2nd edition

GAY Daniel, GAMBELIN Jacques
Modeling and Dimensioning of Structures

MEUNIER Gérard
The Finite Element Method for Electromagnetic Modeling

2005

BENKHALDOUN Fayssal, OUAZAR Driss, RAGHAY Said
Finite Volumes for Complex Applications IV: Problems and Perspectives

Printed and bound by CPI Group (UK) Ltd, Croydon, CR0 4YY
16/09/2026
14963076-0002